# The Logboats of Scotland
*With Notes on Related Artefact types*

Robert J C Mowat

Oxbow Monograph 68
1996

*Published by*
Oxbow Books, Park End Place, Oxford OX1 1HN

© Oxbow Books and the author 1996

ISBN 1 900188 11 2

This book is available direct from
Oxbow Books, Park End Place, Oxford OX1 1HN
*(Phone: 01865-241249; Fax: 01865-794449)*

*and*

The David Brown Book Company
PO Box 5111, Oakville, CT 06779, USA
*(Phone: 860-945-9329; Fax: 860-945-9468*

*Printed in Great Britain by*
Information Press, Oxford

# Contents

*List of Plates*..................................................................................................................................................................*v*
*List of Figures*................................................................................................................................................................*vii*
*List of Tables*.................................................................................................................................................................*ix*

I      Introduction, Methodology and Acknowledgements..........................................................................................1

II     Gazetteer of Logboat Discoveries......................................................................................................................11

III    Examples of Related Artefact-types...................................................................................................................81

IV    Synthesis and Analysis.....................................................................................................................................109

    IV.1          Timber Supply................................................................................................................................109
          IV.1.1     The Northern and Western Islands.................................................................................114
          IV.1.2     The Northern Highlands..................................................................................................114
          IV.1.3     The East-Central Highlands............................................................................................114
          IV.1.4     The West-Central Highlands..........................................................................................115
          IV.1.5     The Central Lowlands, Strathmore, The Borders and the South-West.........................115

    IV.2          Geographical Distribution and the History of Discovery............................................................116
          IV.2.1     Dumfries, Galloway and Ayrshire..................................................................................119
          IV.2.2     Clyde estuary and river-terraces......................................................................................120
          IV.2.3     Highland lochs and bogs (including Argyll, the Trossachs and the Western Isles).......120
          IV.2.4     Strathmore, Angus, the Mearns and the North-east......................................................121
          IV.2.5     Tay estuary......................................................................................................................121
          IV.2.6     Forth valley and estuary..................................................................................................121
          IV.2.7     Miscellaneous .................................................................................................................121

    IV.3          Surviving Examples and Detailed Records..................................................................................122

    IV.4          Structural Features.........................................................................................................................122

    IV.5          Size, Form and Morphology..........................................................................................................123

    IV.6          Repairs ...........................................................................................................................................125
          IV.6.1     Patch nailed or pegged over hole or knot......................................................................126
          IV.6.2     Block or timber inserted into hole or split.....................................................................126
          IV.6.3     Oversewn batten along split...........................................................................................126

    IV.7          Dating and Chronology..................................................................................................................126
          IV.7.1     Radiocarbon dates...........................................................................................................129
          IV.7.2     Archaeological association, including incorporation into crannog structures................129

|  |  |  |
|---|---|---|
| | IV.7.3 | High antiquity indicated on geological grounds............132 |
| | IV.7.4 | Pollen and vegetational dating............133 |
| | IV.7.5 | 'Modern' features............133 |
| | IV.7.6 | Constructional use of iron or unspecified 'metal'............135 |
| | IV.7.7 | Dendrochronology ............135 |
| IV.8 | Paddles and Oars............136 | |
| IV.9 | Miscellaneous Discoveries............137 | |
| | IV.9.1 | Log-coffins............137 |
| | IV.9.2 | Troughs, kegs and bog butter containers............141 |
| | IV.9.3 | Mill-paddles and troughs............144 |
| | IV.9.4 | Cooking- and boiling-troughs............145 |
| | IV.9.5 | Salt-making troughs............147 |
| | IV.9.6 | Retting and industrial troughs............148 |
| | IV.9.7 | Sledges and slides............148 |
| IV.10 | Conclusions ............148 | |
| V | Bibliography and Abbreviations ............151 | |
| Index............163 | | |

# List of Plates

*1. Cambuskenneth (no. 14)*.................................................................................................................................*15*
Apparently-inaccurate but artistically satisfying lithograph impression of the scene at the time of discovery. Reproduced, by kind permission of the publishers, from the *Illustrated London News* of 6 June 1874. (NMRS DC 11554/p).

*2. Dowalton Loch 1 (no. 29)*............................................................................................................................*25*
Left hand: plan and quarter-view of the logboat, probably drawn by Robertson or Percy at the time of discovery, but not included in the published account.
Right-hand: evidently depicts the 'heavy slabs of oak...laid upon one another in a sloping direction, bolted together by stakes inserted in mortises...and connected by square pieces of timber' that secured the upper part of the south side of the crannog (Munro 1885, 86).
Reproduced from a drawing in the NMRS (WGD/49/1/2; SAS 439) by kind permission of the Royal Commission on the Ancient and Historical Monuments of Scotland.

*3. Dumbuck (no. 35)*......................................................................................................................................*27*
Romantic impression by W Milne Black of the crannog and logboat in use, which is included here as a comment upon the presumed carrying capacity of an apparently-small logboat. Reproduced by courtesy of the Trustees of the National Library of Scotland from the *Scots Pictorial* for 29 October 1898.

*4-5. Erskine 6 (no. 44)*...................................................................................................................................*33*
Two general views taken during recovery operations and showing the probable anchor damage. The boat was apparently turned round between the taking of the two photographs. Reproduced by kind permission of Miss H Adamson from the collection of Glasgow Museums and Art Galleries.

*6. Glasgow, Clydehaugh 4 (no. 56) and Glasgow, Clydehaugh 5 (no. 57)*.................................................*37*
These boats are depicted on the left and right sides respectively of a watercolour, dating from 1852, by A McGeorge. This view probably idealises the form of the boats and exaggerates their major dimensional characteristics, but is invaluable for their depiction of the fittings. Reproduced by kind permission of Miss H Adamson from the collection of Glasgow Museums and Art Galleries.

*7. Glasgow, Hutchesontown Bridge (no. 59)*................................................................................................*38*
Bow view of forward section as currently displayed. Reproduced by kind permission of Miss H Adamson from the collection of Glasgow Museums and Art Galleries.

*8. Glasgow, Springfield 1 (no. 64) and 2 (no. 65)*.......................................................................................*42*
These boats are depicted on the left and right sides respectively of a watercolour, dating from 1847, by A McGeorge. Reproduced by kind permission of Miss H Adamson from the collection of Glasgow Museums and Art Galleries.

*9. Lea Shun (no. 85)*......................................................................................................................................*49*
Undated photograph by John B Russell of the remains of this boat, presumably taken at Tankerness House. Reproduced from the manuscript collection of the Society of Antiquaries of Scotland held in the Museum of Scotland (SAS Ms. 307/12). I am grateful to the Society of Antiquaries of Scotland for permission to reproduce this plate.

*10. Loch Arthur 1 (no. 92)*............................................................................................................................*51*
View of logboat, apparently soon after its discovery, and said to have been drawn by Professor Geikie. Reproduced from the manuscript collection of the Society of Antiquaries of Scotland held in the Museum of Scotland (SAS Ms. 307/12). I am grateful to the Society of Antiquaries of Scotland for permission to reproduce this plate.

*11-14. Loch Arthur 1 (no. 92)*.......................................................................................................................*53*
Views and details reproduced by kind permission of the Trustees of the National Museums of Scotland:
11. View of port side, from bow.
12. Detail of port side of 'animal-head' bow.

13. Detail of starboard side of 'animal-head' bow.
14. Detail of stepped interior of forward section, looking forward.

*15-17. Loch Glashan 1 (no. 102)* ............................................................................................................59
Photographs taken at the time of discovery and reproduced from the negatives held in the NMRS (A53503-5) by kind permission of Mr JG Scott and of the Royal Commission on the Ancient and Historical Monuments of Scotland.
15. General view from the port side.
16. View from the port bow, showing the general form of the boat, the transom and the thwart.
17. View from astern.

*18. Loch Laggan 2 (no. 110)* ..................................................................................................................64
General view from the bow, showing evidence of repair. Reproduced from plate xvi (lower) of volume lxxxv (for 1950-1) of the *Proceedings of the Society of Antiquaries of Scotland*. I am grateful to the Society of Antiquaries of Scotland for permission to reproduce this plate.

*19. Loch of Kinnordy (no. 118)* .............................................................................................................66
Publication drawing by Lyell, showing the stylised depiction of the suggested animal-head and the setting of the boat in its geological context. Reproduced, by kind permission of the Geological Society and of Edinburgh University Library, from the *Transactions of the Geological Society of London* for 1829.

*20. Loch Glashan, crannog, paddle (no. A38)* .......................................................................................93
Photograph taken *in situ* during excavation and reproduced from the negative held in the NMRS (A53507) by kind permission of Mr JG Scott and of the Royal Commission on the Ancient and Historical Monuments of Scotland.

*21. Loch Glashan, crannog, paddle-shaped object or 'bat' (no. A40)* ...................................................94
Studio photograph with another timber artifact, the 'bat' being the lower. Reproduced from the negative held in the NMRS (A36978) by kind permission of Mr JG Scott and of the Royal Commission on the Ancient and Historical Monuments of Scotland.

*22. Loch Glashan, crannog, bowls and dishes, including no. A44: upper* ............................................98
Studio photograph reproduced from the negative held in the NMRS (A36975) by kind permission of Mr JG Scott and of the Royal Commission on the Ancient and Historical Monuments of Scotland.

*23. Oakbank, crannog, paddle (no. A56)* ............................................................................................102
Dorsal view reproduced by kind permission of Dr TN Dixon.

# List of Figures

*1. Distribution Map* .................................................................................................................................9
Distribution of logboat discoveries, indicating possible examples, approximate locations, and groups of finds. Scale 1:212,000.

*2. Cambuskenneth (no. 14)* .....................................................................................................................16
Logboat. Plan, starboard elevation and section at 33% of length from bow. Scale 1:25.

*3. Closeburn (no. 21)* ............................................................................................................................19
Logboat. Plan, starboard elevation and section at 12% of length from bow. Scale 1:25.

*4. Craigsglen (no. 23)* ...........................................................................................................................19
Logboat. Plan view of surviving remains. Scale 1:25.

*5. Dalmarnock (no. 25)* .........................................................................................................................23
Logboat. Plan view, and section near one end. Scale 1:25.

*6. Eadarloch (no. 36)* ............................................................................................................................23
Possible logboat. Plan view with section roughly halfway along length. Scale 1:25.

*7. Dumbuck (no. 35)* .............................................................................................................................26
Logboat. Plan view of remains. Scale 1:25.

*8. Errol 2 (no. 38)* ................................................................................................................................29
Logboat. Plan view and port elevations with sections at 36% and 85% of length from bow. Scale 1:25.

*9. Erskine 6 (no. 44)* .............................................................................................................................34
Logboat. Plan, port elevation and four sections at roughly equidistant spacings: re-worked from excavation drawings. Scale 1:25.

*10. Garmouth (no. 51)* ..........................................................................................................................36
Logboat. Plan, starboard elevation and section at 20% of length from bow. Scale 1:25.

*11. Glasgow, Hutchesontown Bridge (no. 59)* ...........................................................................................39
Logboat. Plan, starboard elevation and sections at 45% of length from bow, and at the stern. Scale 1:25.

*12. Glasgow, Rutherglen Bridge (no. 63)* ................................................................................................39
Logboat. Plan and starboard elevation. Scale 1:25.

*13. Glasgow, Springfield 1 (no. 64)* ........................................................................................................42
Logboat. Plan, starboard elevation and section amidships. Scale 1:25.

*14. Glasgow, Springfield 2 (no. 65)* ........................................................................................................43
Logboat. Plan, port elevation and sections amidships and at stern. Scale 1:25.

*15. Glasgow, Springfield 5 (no. 68)* ........................................................................................................43
Logboat. Plan view, starboard elevation and section amidships. Scale 1:25.

*16. Kilbirnie Loch 3 (no. 77)* ................................................................................................................46
Logboat. Plan, elevation of (probable) starboard side and section amidships. Scale 1:25.

*17. Kilbirnie Loch 4 (no. 78)* ................................................................................................................46
Logboat. Plan, starboard elevation and section at roughly midpoint of surviving remains. Scale 1:25.

*18. Loch Arthur 1 (no. 92)* ....................................................................................................................52
Logboat. Plan, starboard elevation and two sections of preserved forward section, with interior elevation and (partial) section of detached stern. Scale 1:25.

*19. Littlehill (no. 90)* ............................................................................................................................56
Logboat. Plan, starboard elevation (as currently retained) and section at 37% of length from the bow. Scale 1:25.

*20. Loch Doon 1 (no. 96)* .....................................................................................................................56
Logboat. Plan, port elevation and section amidships. Scale 1:25.

*21. Loch Doon 3 (no. 98)*..................................................................................................................................57
Logboat. Plan and port elevation with sections at 77% of length from bow, and at stern. Scale 1:25.

*22. Loch Glashan 1 (no. 102)*..............................................................................................................61
Logboat. Plan, starboard elevation and section midships. Scale 1:25.

*23. Loch Leven (no. 116)*....................................................................................................................61
Logboat. Plan, elevation and section of possible surviving remains of possible logboat. Scale 1:25.

*24. Loch of Kinnordy (no. 118)*..........................................................................................................67
Logboat. Plan and port elevation of surviving remains, with section amidships. Scale 1:25.

*25. Lochlea 3 (no. 126)*......................................................................................................................67
Logboat. Plan, elevation of (probable) starboard side and section amidships. Scale 1:25.

*26. Lochmaben, Castle Loch 2 (no. 131)*............................................................................................71
Logboat. Plan of surviving remains, as currently retained. Scale 1:25.

*27. Lochmaben, Kirk Loch 1 (no. 132)*..............................................................................................71
Logboat. Plan of surviving remains. Scale 1:25.

*28. Lochmaben, Kirk Loch 2 (no. 133)*..............................................................................................71
Logboat. Plan of surviving remains. Scale 1:25.

*29. 'Orkney' (no. 140)*........................................................................................................................75
Logboat. Plan and starboard elevation of largest (divided) portion with interior elevation of detached fragment of stern, exterior elevation of transom, and sections on either side of division in the main portion and at about 70% of the (surviving) length from the bow. Scale 1:25.

*30. 'River Clyde' (no. 149)*................................................................................................................77
Logboat. Plan, port elevation and section at 34% of length from the bow. Scale 1:25.

*31. White Loch (no. 154)*....................................................................................................................79
Logboat. Plan view of surviving remains and section near bow. Scale 1:25.

*32. Daviot, timber (no. A15)*..............................................................................................................79
Unworked timber, formerly proposed as possible logboat. Plan view, for comparison with accepted logboat remains. Scale 1:25.

*33. Glenfield, Kilmarnock, ard (no. A25)*........................................................................................88
Plan view with section of blade. Scale 1:10.

*34. Loch Glashan, paddle or oar (possible) (no. A36)*....................................................................88
Plan of dorsal surface with sections at each end of surviving portion. Scale 1:10.

*35. Lochlea, crannog, oar (no. A51)*................................................................................................88
Plan of blade and part of loom, with sections of distal end of blade and lower part of loom. Scale 1:10.

*36. Tentsmuir, paddle (no. A69).*
Plan view of dorsal side with sections of distal end of blade and at three points along loom and handle. Scale 1:10.

*37. Lochlea, crannog, 'double-paddle' (no. A52)*............................................................................88
Plan view of surviving portion, with section of (broken) blade, and at one end of loom. Scale 1:10.

*38. Loch Glashan, crannog, model paddle (possible) (no. A41)*....................................................95
Plan view at scale 1:2.

*39. Loch Glashan, crannog, paddle (no. A39)*................................................................................95
Plan of dorsal side. Scale 1:10.

*40. Loch Kinord, paddle (no. A46)*..................................................................................................95
Plan view of surviving portion, with section of lower loom. Scale 1:10.

*41. Ravenstone Moss, paddle (no. A62)*..........................................................................................95
Plan view of surviving portion with section of lower loom. Scale 1:10.

*42. Rubh' an Dunain, Skye, oar or paddle (possible) (no. A65)*....................................................95
Plan view of surviving portion, with section of lower loom. Scale 1:10.

*43. Eadarloch, trough (no. A18)*......................................................................................................97
Plan and elevation with section at midpoint. Scale 1:25.

*44. Loch Glashan, crannog, trough 1 (no. A42)*..............................................................................97
Plan and elevation with section (at midpoint). Scale 1:25.

*45. Loch Glashan, crannog, trough 2 (no. A43)*..............................................................................97
Plan and elevation with section (at midpoint). Scale 1:25.

*46. Loch Glashan, crannog, trough 3 (no. A44)*..............................................................................97
Plan and elevation with section (at midpoint). Scale 1:25.

# List of Tables

1. Concordance table with Edwards numbers..................110
2. Comparative figures for states of logboat preservation in Scotland as against England and Wales, in accordance with the criteria defined in section IV.3..................123
3. Comparative figures for logboat preservation in specific areas of Scotland..................124
4. Comparative figures for state of surviving logboats in specific areas of Scotland..................125
5. Figures for types of timber identified in Scottish logboats with comparable figures for England and Wales..........126
6. Figures for recorded lengths of Scottish logboats with comparable figures for England and Wales..................127
7. Figures for identifiable forms of Scottish logboats with comparable figures for England and Wales..................127
8. Summary of recorded radiocarbon dates for European logboats, in reverse chronological order..................130
9. Summary of sizes and forms of Scottish oars and paddles..................138
10. Summary of circumstances of discovery of domestic wooden vessels and similar artefacts in Scotland..................141
11. Summary of the forms, dimensions and capacities of bog butter troughs and kegs, the mill lade from Cairnside and selected troughs, bowls and dishes of probable domestic origin being included for comparison. All dimensions are cited in metres, except for the capacity which is in litres. Internal measurements (where available) are cited beneath the relevant external figure..................145
12. Comparative summary of the numbers of artefacts of various type from the Loch Glashan crannog and from the Iron Age lake villlage at Glastonbury, Somerset (after Earwood 1988)..................147

# I Introduction, Methodology and Acknowledgements

> What coms't thou to reveal - thou battered ark,
> From out thy sandy bed so rudely hurried,
> Will thine appearance serve to throw one spark
> Of light upon the age when thou wert buried?
> 'To the old canoe of the Clyde'
> Stuart 1848, 48.

The study of logboats was considered of major importance during the formative years of Scottish archaeology, and accounts of their discovery were enthusiastically collated by Munro, Stuart and other Victorian antiquaries largely on account of their supposed value as chronological indicators. In the early part of the present century they came to be seen as of little significance, and received correspondingly less attention. In post-war times, however, further new discoveries (few of them adequately published), an increasing interest in dendrochronology, and the advent of radiocarbon dating, together with the compilation of three (unpublished) analytical studies have served to revive interest in the subject.

During the early 1950's Scottish logboats were studied by Mrs E. Grant of Edinburgh University Archaeology Department. The published discoveries were collated and the surviving remains drawn for a lecture delivered to the Society of Antiquaries of Scotland on 10th November 1952, but were not published. In 1966, (Dr.) J.M. Graham, then of the University College of South Wales, Cardiff, studied the English and Welsh boats for an undergraduate dissertation but included some account of the Scottish examples; again, this work was not published. Finally, in about 1981, Miss Denny Edwards, also in an undergraduate dissertation written from Cardiff, summarised the Scottish examples; her untimely death precluded publication but a concordance of the present work with hers is appended (table 1).

It thus falls to the present author to publish the first corpus of Scottish logboats since Stuart (1866a, 148–51, 174).

In so doing, full account has been taken of the methodology and terminology established for this class of artefact by McGrail (1978, i, 26–102), to which work this study may be taken as a sequel. In particular, the same criteria are used to distinguish logboats from other similar timber artefacts as have been laid down by McGrail (1978, i, 19, 93; 1987a, 57) in addressing the problem of distinguishing logboats from other similar timber artefacts. He proposes that an artefact of monoxylous dugout be accepted as a logboat if it satisfies two or more of the following conditions:

1. it is found in or near a (former) watercourse.

2. it is associated with other nautical artefacts, such as fishing tackle, anchors, paddles and poles.

3. at least one end is shaped in one of the well-documented logboat forms.

4. it has (vestigial) fittings which are appropriate to logboats, such as thwarts, thole-pins, ribs and stabilisers.

5. it measures more than about 3m in length.

6. the bark and sapwood have been removed.

The advantages and disadvantages of monoxylous boat construction and use have been discussed at length by Greenhill (1976, 129–52 *passim*), McGrail (1978, i, 28–51 *passim*, 88–93, 95–102, 117–25; 1987a, 56–87 *passim*) and McKee (1983, 47–8, 55–9). McKee has particularly stressed both the essential simplicity of the logboat or dugout concept and the concurrent disadvantages from which it suffers on account of its propensity to splitting at the ends. Compared to the unhollowed floating log, which has no inherent stability and can only roll uncontrollably, the hollowed boat offers benefits in both stability and carrying capacity, as the higher-density payload serves to retain the centre of gravity below that of buoyancy.

On the basis of the consideration of the artefact-types discussed above, it is suggested that the presence or absence of thickness-gauge holes (McGrail 1978, i, 31–2) should be considered a particular indicator of the classification into which such an object should be placed. The drilling of such features forms a major part of the standard method of logboat manufacture (McGrail 1987a, 59–64), and they are well-attested both archaeologically and ethnographically from localities as diverse as Europe, the Indian sub-continent, the Far East and the Amazon basin.

The presence or absence of such holes can readily be determined even from incomplete specimens and poor accounts of lost examples, while their existence greatly facilitates the hollowing of a trough to sufficiently close tolerances to serve as a well-trimmed logboat of adequate capacity. Poor or non-provision of these aids may result in the construction of a misshapen, unseaworthy and inefficient craft, whereas poor workmanship in the case of any of the other categories of monoxylous artefact will not have such damaging effects on the utility of the finished object.

The function of these holes has frequently been misunderstood. In an Irish context, Macalister (1928, 192) mentions the discovery of a logboat in Lough Erne, Co. Fermanagh, the 'keel' of which was pierced by nine holes which were presumably intended to retain thickness-gauges. He memorably misinterpreted these as evidence that 'some delinquent had been punished by being set adrift in a leaky boat'.

That these basic definitions remain susceptible of amendment over a century after the beginning of systematic study indicates the fundamental problems posed by a type of artefact which appears simple but has often proved as difficult of recognition in the field as it has of preservation in the museum.

The reductive technique of manufacture that is definitive of the type is so simple as to have been self-evidently an oft-repeated re-invention. The type itself is thus without cultural or chronological significance, and can no longer be seen as the basis for the development of the planked boat. In spite of this, the wide variation in scantlings, sub-types and fittings seen in Scotland makes the type worthy of study in its own right; future discoveries should be valued accordingly.

To this extent, the logboat may be considered within the category of 'neo-archaic' objects, 'rude implements' or (in modern theoretical terminology) ethnographic comparanda that were studied and recognised by Mitchell (1880, v, 1–13, 96) as then being widespread in those more remote areas of Scotland that he saw as not having advanced to the 'higher culture' of the more developed areas of the East and South. He saw this distinctive material culture as elucidating the modes of manufacture and use of such common archaeological discoveries as spindle-whorls, quernstones, sinkers and fishing implements, various types of agricultural implement, wheelless vehicles, horizontal or 'Norse' mills, 'black' or 'beehive' houses and the occupation of caves as well as the potential for the extempore use of such common objects as animal skins and root vegetables. Mitchell's attitudes may be summed up (1880, 12 and 14n. respectively) in the statements that 'my witnesses shall be chosen from objects and practices in the midst of which we live' and 'primitive is a word of uncertain meaning'.

In the process, he discusses the triple problems of imported artefacts, chronological imprecision and the then-dominant law of progress through natural selection. He notes (1880, 113) that 'It is desirable to go still further in showing how this (three-age) classification is defective when it is regarded as marking necessarily successive steps of progress, and to ask whether it is not difficult to see why a man who uses bronze weapons should be inferior either in culture or capacity to a man who uses iron weapons', while also considering (1880, 117) that 'a classification of antiquities into those belonging to the stone, bronze and iron ages, has no absolute chronological significance, and does not furnish dates'. His comments on the archaeological remains likely to survive and to be recovered from specific ethnographic activities are fully worthy of the approach of a modern theoretical or experimental archaeologist, while his analysis of the inter-relationship of prehistoric burial-cairns, modern funerary or 'coffin' cairns and 'beehive' houses can only be described as precocious, even if long superseded. In recent years, the development and widespread use of scientific methods of absolute dating have served to remove this problem, as is graphically illustrated by the radiocarbon-based confirmation of the traditional attribution to the Early Bronze Age (on the basis of ceramic similarities) of the carved wooden polypod bowls of central Ireland (Earwood 1992).

This approach also formed part of the intellectual background of Joseph Anderson, the greatest Scottish archaeological systematiser of the Victorian and Edwardian periods, who (from a viewpoint of a museum appointment) set the study of prehistoric artefacts in a more general archaeological context in two major studies (1883, 1886). His recognition of the significance of the polished stone axe from Solway Moss that was found in the dried remains of its handle and his publication of it with comparative examples from the ethnography of the Pacific basin (1886, 351–4) is a simple example of the application of this methodology to the material assemblage of distinctive quality that is so frequently found in Scotland. Being more concerned to establish a chronological sequence (after the manner of Worsaae), he omits the study of logboats and monoxylous coffins, but pays great attention to such varied classes and types as Viking ship-burials, burnt mounds and textiles (1883, 62–5, 72–3 and 102–6 respectively).

Munro, the great explorer of crannogs and lake dwellings, was himself less concerned with this form of analysis although his extended studies (1897, 239–307) of animal traps and bone skates from non-Scottish contexts demonstrates that he was well aware of it.

More recently, the same approach has been used by Fenton to demonstrate the range of possible human technological responses to a broadly-comparable environment, most notably in his study of land transport (1984a, b). He cites evidence for the use in highland and northern Scotland of a wide range of methods (far more than are now evident) including human porterage, split-saddles, crook-saddles, wooden panniers and basketry creels and kishies for horse transport, and varied forms of travois or slide-car (pulled by either humans or animals) as well as single poles. As an aside, it is immediately apparent that the willow that was specially grown (1984a, 107) must be seen as also forming the potential basis for skinboat construction. In summary, Fenton lays down (1984a, 105) that 'Every piece of evidence has its part to play in building up a science with rules and theories and general concepts. Detailed local studies, 'point' studies are the very stuff of its existence, provided they are put into perspective through wider, collaborative activity'. It is the distinctive nature of Scottish archaeology that a wide range of analytical techniques can be brought to bear on a raw material assemblage of unusual quality and quantity.

The core of this present work (Section II) comprises a gazetteer of logboat discoveries which is arranged in alphabetical order of site-names and loosely follows the conventions of the Archaeological Sites and Monuments series of publications of the Royal Commission on the Ancient and Historical Monuments of Scotland. Each entry is prefaced by a serial number, site-name, National Grid Reference and 1:10,000 or 1:10,560 map number followed by the serial number of the appropriate NMRS dataset. To this is appended a summary of the administrative area or areas within which the site is deemed to lie. Both the appropriate parishes and 'traditional' (post-1890) counties and (post-1975) Regions and Districts are cited; the latter have been found to provide a valuable framework of suitably-sized areas for geographical analysis, although it is appreciated that they have themselves since been superseded.

The text of each entry describes the location, geographical setting and circumstances of discovery of each boat, and summarises any surviving record of the vessel as found. Details of the present location and condition of any remains and of any known scientific analysis, dating, restoration or treatment are added whenever appropriate while reference is made to any explanatory drawings and photographs. All radiocarbon dates cited as 'bc' are believed to be uncalibrated and based on the 5570-year 'old' half-life.

Scottish archaeology is distinguished by being based on a large (if under-exploited) reservoir of archaeological potential in extensive deposits of peat and wetland, and in having a wealth of comparative information in the distinctive 'native' or 'traditional' material culture that has formed the subject of ethnographic study since a relatively early date. This potential is sometimes shared and often directly comparable with that of Ireland and (to a lesser extent) Wales and England. Any study of Scottish logboats must consider both these circumstances.

The extensive peat cover (of all types) that is typical of large areas of both highland and lowland Scotland (Coppock 1976, 13, fig. 4; Earwood 1993a, 4, fig. 2) were early shown to contain large quantities of organic archaeological artefacts, including textiles and clothing, figurines, timber trackways, bog butter troughs and kegs, and the remains of wheeled vehicles. Most of the numerous discoveries that were made during the agricultural improvement phase of the later eighteenth century must have been lost without record, and it is thanks to the enthusiastic early members of the Society of Antiquaries of Scotland that a (hopefully representative) sample survives in the collections of the Royal Museum of Scotland as a basis for research. Further agricultural, forestry and (particularly) industrial development in the nineteenth and twentieth centuries have added significantly to the collections of the Royal and other museums. In particular, the systematic excavation (in 1960) of the crannog in Loch Glashan, Mid-Argyll (RCAHMS 1988, 205–8, no. 354) was a notable advance as it yielded the largest stratified assemblage of organic artefacts yet found in Scotland; following conservation, these objects are held in Glasgow Art Gallery and Museum. Since then, road and motorway construction (particularly in southern Scotland) has necessitated the stripping of large areas of peat, with numerous resulting discoveries, the incidence of which is probably related to the degree of locally-available archaeological supervision.

The numerous eighteenth-century 'travellers' (of whom Johnson and Boswell are only the most famous) recorded a distinctive society, language and material culture in the north and west while (to a lesser degree) distinctive local traditions were present (and, indeed, still developing) in other regions. Roughly a century later, Mitchell (1880) attempted to use such comparative material when interpreting aspects of Scottish archaeology in the light of his 'neo-archaic objects' and within the context of the then-prevalent concepts of natural selection, inevitable progress and the three-age system.

He was keen to demonstrate (although not in these terms) the conserving nature of highland society, and was probably far out of step with the orthodoxy of his day in rejecting the application of at least the two last-cited principals in the context of his study, stressing instead the continuity of such artefact-types as spindle-whorls of clay, stone and steatite, and recognising the potential use in antiquity of archaeologically-unrecognisable practices as cooking in a skin. He argues (1880, 113) that 'It is desirable to go still further in showing how this classification is defective when it is regarded as marking necessarily successive steps of progress, and to ask whether it is not difficult to see why a man who uses bronze weapons should be inferior either in culture or capacity to a man who uses iron weapons' while elsewhere (1880, 14) he categorically states that 'primitive is a word of uncertain meaning'.

In his study of then-current 'objects and practices' he comes close to the Childean concept of material culture, specifically contrasting (1880, 25–32) the locally-made and the imported industrial products then to be found in typical house in terms of the remains he expected to be found following the destruction and excavation of the building, an idea fully worthy of any modern experimental or theoretical archaeologist.

Most important of all, Mitchell (1880, 117) recognises the problem of absolute chronology and the limitations of the then-limited dating techniques. In the absence of detailed studies of peat stratigraphy and scientific dating, many of this 'neo-archaic' objects (particularly those from most peat and wetland deposits) are effectively unstratified and, given the traditional technological conservatism of much of Scotland, he recognises that a 'classification of antiquities into those belonging to the stone, bronze and iron ages, has no absolute chronological significance, and does not furnish dates'. A further century on, Piggott (1983, 26) was to point out that the then-undated disc-wheel from Blair Drummond Moss, Perthshire 'could equally be of the eighteenth century AD or BC'; it has since been dated to 860 ± 85 bc (OxA-3538), but the point remains a valid one, and could be applied to any of a wide variety of Scottish artefact-types.

This form of analysis has been applied by Fenton (in particular) to both wheeled and wheelless forms of land transport (1984a and b). He seeks to demonstrate a 'constant interaction between community, environment, equipment and sometimes ways of doing things that involve no equipment' (1984a, 105).

In principle, the distinctive properties of the logboat appear well-suited for use over much of Scotland as recently summarised by Martin (1992). The heavily-indented coastline of the western seaboard and its islands comprises a plethora of sheltered bays and inlets (of widely-varying sizes) where such vessels might be used in a reasonable degree of safety within easy reach of shore. Inland, a multitude of lochs, lochans, marshes and mosses (many of them small and now-drained) would have offered conditions for logboat use only surpassed for suitability by the English Fenland, the Somerset Levels and the loughs of the Irish midlands. Socially, the presumed small and scattered population would, in general terms, appear a suitable basis for the operation of a type of vessel well suited for construction and use by non-specialists. In such contexts, the logboat may be envisaged as one of a number of predecessors to the fibreglass workboats that are used for fishing and light transport in such contexts today.

Against this must be set the major limitation that the maximum size of any logboat is determined by that of the parent log. Although no figures are apparently available for the size and shape of trees and timbers in pre-industrial Scotland, it must be assumed that (before the recent introduction of specialist mass commercial species) most trees growing in southern and central Scotland were near to, or on the edge of, their natural distribution, and were accordingly of smaller size than might be expected under more favourable conditions. Further north, the Scottish landscape has been either treeless or naturally forested by coniferous trees over several millennia, and the raw material for logboat construction was accordingly highly restricted.

In this context, the efficiency of the logboat in terms of the proportion of raw material removed is reduced by the necessity to leave sufficient spare timber to prevent splitting, particularly if inadvertent drying-out should occur or if holes are to be drilled into the wood or fittings nailed into position. End-forms developed for this reason include duck-bill or platformed ends and those rounded in plan, section or both. The number of preserved Scottish examples that have been reduced by splitting rather than abrasion to little more than flat planks testifies that the importance of providing a generous radius of curvature to internal corners and of providing ample margins of separation for the 'safe nail' line (so as to avoid splitting) was not appreciated by many Scottish logboat builders.

For this reason, the dangers should be stressed of employing any more sophisticated form of quantitative analysis than is justified by the nature of either the evidence or the subject, and such an approach eschewed. The surviving remains of the Scottish examples are frequently distorted by shrinkage and warping and are universally incomplete; in almost all cases the sides of the vessel have not survived to their full height and the sheerline cannot be established with any certainty, if at all. Unfortunately, the increased risk of damage associated with boats with flared sides inevitably results in an under-representation of such craft among the better-preserved examples. The same is true of well-built boats constructed to thin scantlings and with a sharp angle at the junction of the side with the floor; such vessels stand a high chance of being reduced to a flat plank. It is in the nature of the evidence that the logboats most likely to survive to preservation are small, thick-sided and crudely-built examples of rounded form. The picture of logboat construction presented is thus biased towards the less sophisticated examples.

This concern is reinforced if it is accepted that the logboat was a simple and unsophisticated type of craft which could be built with relative ease from available raw materials and with a restricted tool-kit by non-specialist builders who did not have the thorough theoretical and empirical understanding necessary to build a skin or plank boat. Detailed analysis of building traits and methods of utilisation implies that these characteristics were understood by the builders and users themselves. In short, the use to which a logboat was put may have been more a function of those capabilities demonstrated after launch than those intended beforehand.

In her study of domestic wooden artefacts, Earwood (1993, 234–43) identifies a four-stage theoretical progression from the manufacture of single-piece objects, through that of (generally larger) artefacts of jointed, sewn or bentwood construction to the manufacture of complex plank-and stave-built artefacts, and lathe-turned work. She considers (1993, 151, 154, 159, 243–4) the manufacture of logboats to be essentially the same as that of any other type of trough, with no discernible variation in technique over time, although differences in style and decoration may be recognised. The construction of two-piece kegs is considered comparable to that of logboats with detached transoms, and the sewn or laced construction of two-piece containers is comparable to that used in the construction of sewn boats and, on occasion, in the repair of logboats.

However, the practical design and construction of logboats is complicated by the necessity to provide an adequate degree of stability and possibly also by a desire to achieve a specific level of performance in the fulfilment of a specific task under specific conditions. In the archaeological analysis of such characteristics (whether by reconstruction or computer simulation), a major difficulty is the necessary assumption that no water has entered the boat and that no ballast was carried. In practice, the small size and low freeboard characteristic of logboats must have made a completely dry boat a highly unusual phenomenon, particularly in the adverse conditions under which stability is a paramount requirement. The proposed identification by Andersen (1986, 94–5) of the flat-sided stone of 30kg weight that was found in the Tybrind Vig I logboat as ballast is also significant in this context.

Accordingly, calculations are made in the case of each boat for which adequate measurements can be derived so as to ascertain the potential uses of each in general terms. The method used is a variant of that devised by McGrail (1978, i, 94–102, 134–42 and 304–49 *passim*; 1987a, 12–22, 192–203; 1988; *pers. comm.*) but it is stressed that the results are essentially relative, there being little comparative data. That they serve best to illustrate the range of possibilities is suggested by the comparative drawings of Rieck and Crumlin-Pedersen (1988, 85) that indicate the varying combinations of crew and cargo that might be fitted within specimen sets of parameters.

The following indices are defined for this purpose:

1. Slenderness coefficient
(McGrail 1987a, 194, 197)
This is defined as overall length divided by greatest beam, and is an indicator of speed potential. Narrow boats (taken as having a slenderness coefficient of 3.75 or higher) are fast and easy of propulsion. Those with a slenderness coefficient of 2.6 or less are described as beamy and tend to be directionally stable, although this is dependent on the depth of draught and, consequently, upon their lading.

2. Beam/draught coefficient
(McGrail 1978, i, 139)
This is defined as beam divided by draught and is an indicator of the type of cargo carried. Boats with higher volumes are 'volume dominated' and so are adapted for the carriage of low-density (bulky) cargoes such as grass or dry peat.

To illustrate the range of densities for various cargoes that might reasonably be expected to have been carried in logboats, the following values (in kg/cu. m.) have been cited (Millett and McGrail 1987a, 133):

| | |
|---|---|
| Iron ore or stone | 2500 |
| Oak timbers | 800 |
| Grain | 680 |
| Meat joints | 680 |
| Turf or peat | 435 |

3. Block coefficient
(McGrail 1978, i, 138)
This coefficient is the value of the displaced volume relative to that of a rectangular block of the same maximum dimensions, and is calculated by dividing the displacement volume by (length x beam x draught). Boats with the maximum value of 1 are rectangular in shape and section. Some inaccuracy will arise if, as is the case for most logboats, the greatest beam is found above the waterline.

4. Log conversion percentage
(McGrail 1978, i, 311–12)
This is a percentage measure of the amount of timber removed in the construction of a logboat, and hence of the efficiency of that process. It is calculated by taking the volume of the parent log, less that of the remaining timber, divided by the volume of the parent log and multiplied by 10. This percentage may also be expressed as the conversion coefficient, which is calculated by dividing the volume of the parent log into formula into the volume of timber remaining. A high figure may indicate that the boat is carrying significant excess weight, to the detriment of performance and possibly stability, but the figure must be read in conjunction with the load space coefficient to assess the suitability of the vessel as a carrying vehicle.

This calculation assumes that the logboat was built to the greatest possible size from a given log. No allowance is made for sapwood or bark and inaccuracies may result from sectional irregularities in the boat or parent log, or from the size and form of the bow and stern fittings. Figures are cited in terms of reduction from a whole log. In the case of boats apparently worked from a half-log, the percentage is re-calculated on that basis and appended.

5. Load space coefficient
(McGrail 1978, i, 140)
This coefficient compares the volume of the load space hollowed out with that of the parent log and is accordingly a quantitative indication of the efficiency of construction in terms of the volume available for payload. It thus differs slightly but significantly from the log conversion percentage, and is calculated by the dividing the volume of the logboat hollow by the volume of the parent log.

6. Displacement volume
(McGrail 1987a, 199–203; *pers. comm.*)
This figure is the volume of water displaced by the hull of a boat in a given condition and is expressed in cubic metres. It is standardised at the point when the waterline is at 60% of the total depth of the boat, that is to say when the draught divided by the draught plus the freeboard equals 0.6.

7. Seaworthiness coefficient
(McGrail 1987a, 194)
A measure of the safety of flat-bottomed boats in rough water is given by multiplying the depth of hull by the tangent of the angle of flare and dividing the product by the breadth of the bottom, where the angle of flare is the angle in the transverse plane made by the sides with the vertical.

8. Volumetric coefficient
(McGrail 1987a, 197, 198)
A measure of the speed potential of a boat is given by dividing the displacement volume by the cube of the waterline length. Any value less than 0.0001 indicates that the hull has a high speed potential.

9. Midships coefficient
(McGrail 1987a, 197, 198; *pers. comm.*)
The same quality may be assessed by dividing the maximum underwater cross-section by the product of multiplying the beam and the draught. A value of less than about 0.85 indicates a considerable speed potential.

In describing the form of boats, the location of fittings and features is indicated in terms of their distance from the certain or assumed bow as a percentage of the total length of the boat.

The McGrail morphology code (McGrail 1978, i, 129–30 and ii, fig. 205) is added whenever possible and the morphological category of the boat is assessed, as far as possible, in accordance with those defined by McGrail (1978, i, 322–5), provisional and variant forms being noted wherever appropriate. The following forms are defined accordingly:

1. Canoe form
The ends of the boat are rounded in all three planes and the body is tapered with a rounded transverse section. The applicable morphology codes 2/3 2 2 : 222 : 2/3 2 2 are defined for English and Welsh examples.

2. Punt/barge form
The ends of the boat are rectangular in plan and inclined in elevation, while the parallel-sided body is flared or rectangular in cross-section. A wide variety of morphology codes is applicable, namely 1/5 3/5 1/3/4 : 1 1 1/3/4 : 1/5 3/5 1/3/4, all of which have been cited for English and Welsh examples.

3. Dissimilar-ended form
The bow of the boat is rounded in all three planes and the body is tapered with a rounded or rectangular cross-section, while the stern is of (solid) rectangular or (detached) transom form. The morphology codes cited as applicable to English and Welsh examples comprise 1/4/5 1/4 1/2/4 : 2 2 1/2/4 : 2/3 2 2.

4. Box form
All elements of the boat are rectangular in all three planes. The morphology codes 1/5 1 1/4 : 1 1 1/4 : 1/5 1 1/4 are considered applicable to English and Welsh examples.

Finally, the relevant figure and plate numbers are cited wherever appropriate and a date of inspection is appended for surviving examples. Bibliographic references indicate the primary and major secondary sources appropriate to each discovery.

Section III attempts to provide a similar level of description for examples of a variety of related artefact-types, including paddles, probable log-coffins and other recorded primitive boats, in an attempt to supplement those provided by McGrail (1978, i, 18–19). Dimensions are cited and volumes calculated where appropriate, but quantitative analysis is not attempted. This section is not intended to describe all the timber objects that have been found in Scottish underwater, wetland or excavation contexts, but rather to indicate the range and quality of the surviving remains, the difficulties inherent in their recognition and classification, and the range of techniques that have been brought to bear upon their study. For this reason, several varied mis-identified timbers are considered in some detail and several boats which are of types other than logboats but have been found with or near such craft are listed to clarify the record. Inevitably, more detailed consideration is given to objects which are apparently or possibly functionally related to logboats (notably the various discoveries of 'paddles') and to those which have been dated (whether by association, stratigraphy or scientific means) and so possibly give some indication of the chronological range of the use of specific tools, techniques and timber types.

In Section IV, the typological and structural characteristics of Scottish logboats and related artefacts are considered, in comparison with their English counterparts wherever possible, and an attempt is made to analyse their distribution in place and time. This synthesis is subdivided as follows:

IV.1 Timber Supply. This sub-section aims to summarise the available evidence for the timber supply and availability of various types of raw material in Scotland.

IV.2 Geographical Distribution and the History of Discovery. This sub-section considers the geographical distribution of logboat discoveries with regard to the circumstances of their discovery and the distributions of comparable and related classes of finds and monuments, notably crannogs, island-dwellings and fortifications, and votive deposits.

IV.3 Surviving Examples and Detailed Records. This sub-section comprises a summary of the number and identities of the Scottish logboats that survive in whole or in part, have been recorded in detail, or for which reliable accounts exist.

IV.4 Structural Features. This sub-section considers the major structural features recorded with particular reference to those that may be skewomorphic elements derived from plank-built boats.

IV.5 Size, Form and Morphology. This sub-section compares the dimensions, features and morphology codes of Scottish logboats with those noted for English and European examples.

IV.6 Repairs. This sub-section summarises the evidence for repairs to specific logboats.

IV.7 Dating and Chronology. This sub-section summarises the available dating evidence for Scottish logboats.

IV.8 Paddles and Oars. Summary notes.

IV.9 Miscellaneous Discoveries. Summary notes.

IV.10 Conclusions. This sub-section provides suggestions for further research.

Section V comprises the bibliography and abbreviations, and lists the works cited as sources, together with other relevant accounts. The various abbreviations used are expanded, and correspondence addresses are cited for relevant institutions. An index and lists of plates and figures complete the work.

The manuscript notes and working material used in the preparation of this work have been deposited in the NMRS.

The author is deeply and specifically appreciative of the assistance given by:
Miss H. Adamson (GAGM), for making available the results of the recovery operations at Erskine,
Mr. D. Brown and Mr. S. Tribe at Oxbow, for attending to the problems of computer incompatibility and the preparation of a dense and complex text,
Dr. A. Crone (Edinburgh), for making available details of her excavations at Buiston,
Miss C. Earwood (Exeter), for making available the results of her work on the Loch Glashan material in advance of publication,
Mrs E. Grant (Edinburgh), for making available the unpublished results of her work and allowing it to be deposited in the NMRS,
Mr. N. Gregory (Dublin), for undertaking field visits and underwater survey,
Dr. J.N. Lanting (Rijksuniversitet Groningen, Netherlands) for providing radiocarbon dating assistance,
Prof. S. McGrail (Southampton), for advice and assistance (particularly in matters of terminology and metrical analysis) and for making available a copy of the notes and draft articles written by Miss Edwards (the value of which is itself acknowledged),
Miss E.M. Scott (then of the Archaeology Branch, Ordnance Survey, Edinburgh) for compiling a preliminary listing of the examples then recorded,
Mr. J.G. Scott (Glasgow), for making available the results of his excavations on the Loch Glashan crannog.

He particularly wishes to express his thanks to the Trustees of the Keith Muckelroy Memorial Fund for supporting the employment of Mr. J. Borland to prepare the drawings for publication with characteristic quiet efficiency, and to Mr. Borland himself for many helpful suggestions.

He also wishes to thank the publishers of the *Illustrated London News*, The Royal Commissioners on the Ancient and Historical Monuments of Scotland, the Trustees of the National Museum of Scotland, Miss H. Adamson, the Society of Antiquaries of Scotland, the Trustees of the National Museums of Scotland, Mr. J.G. Scott, the Librarian of Edinburgh University and the council of the Geological Society for permission to reproduce illustrations, as individually specified in the list of plates.

To name names is always invidious, and the risk of inadvertent omission is ever-present, but the assistance given by the following comes readily to mind:
in Aberdeen and Grampian: Dr. D. McArdle, Dr. K.J. Edwards, Mr. C. Hunt, Mr. J. Inglis, Mr. G. McDonald, Mrs C. Sangster, Mrs A.N. Shepherd, Mr. I.A.G. Shepherd, Ms. J. Stones;
in Dumfries: Mr. D. Lockwood, Miss C. Ratchford, Mr. A.E. Truckell;
in Dundee: Miss L.M. Thoms, Mr. A. Zealand;
in Edinburgh: Dr. M. Armstrong, Mr. C. Aston, Mr. J. Barber, Mr. P. Dale, Mr. P. Hill, Dr. C. Mills, Dr. I.A. Morrison, Mr. I.W. Morrison, Dr. I.B.M. Ralston, Dr. J.A. Sheridan, Dr. D.G. Sutherland, Dr. T.F. Watkins, Dr. C. Wickham-Jones;
in Forfar: Mr. I. Neil;
in Fort Augustus: Mr. J.A. Grieve;
in Fort William: Miss S. Archibald, Miss S, Marwick;
at GAGM: Miss H. Adamson, Dr. C. Batey, Mr. A. Foxon;
in Glasgow: Prof. L. Alcock, Mr. W.S. Hanson;
in Hawick: Mrs. R. Capper, Ms. E. Hume;
at HS: Mr. P. Ashmore, Mrs L. Linge;
at HM: Dr. E. MacKie, Mr. L. Keppie;
in Inverness and Ross-shire: Mrs P.E. Durham, Miss J. Harden;

in Kilmarnock: Mr. J. Hunter, Mr. C. Woodward;
in Kirkcudbrightshire: Mr. T. Colin;
in Montrose: Mr. N. Atkinson;
in Orkney: Mrs A. Brundle, Dr. R.G. Lamb;
in Paisley: Mrs V. Riley;
in Perth: Mr. M. Hall, the late Mrs M.E.C. Stewart;
at RCAHMS: Mrs C. Allan, Mrs C. Appleby, Mr. P. Corser, Miss C.H. Cruft, Mr. J.L. Davidson, Mr. J.G. Dunbar, Miss T.M. Duncan, Mr. D. Easton, Mrs L.M. Ferguson, Mr. I. Fisher, Mr. I.F.C. Fleming, Mr. S.P. Halliday, Miss M. McDonald, Mr. P. McKeague, Mr. A. Maclaren, Mr. K. Mclaren, Mrs A.P. Martin, Mr. G.S. Maxwell, Mr. R.J. Mercer, Mrs R. Moloney, Mrs D.M. Murray, Mrs R. Nichol, Dr. J.N.G. Ritchie, Miss A.L. Salmond, Mr. J. Sheriff, Mr. D.A. Smart, the late Dr. I.M. Smith, Mr. G.P. Stell, Mr. J.B. Stevenson;
at RMS: Mr. T. Cowie, Dr. A. Fenton, Mr. C. Hendry, Mrs J. Moran, Mrs A. O'Connor, Mr. A. Quinn, Mr. I. Scott, Dr. J.A. Sheridan, Mr. T. Skinner, Mr. C. Wallace, Mr. T. Ward;
in St. Andrews: Mrs E. Proudfoot;
at SIMS: Mr. M. Dean, Miss D. Groome, Mr. R. Lawrence, Dr. C. Martin, Mr. I. Oxley, Dr. R. Prescott, Miss A. Walker;
in Stirling: Mr. E.J. Fereday, Mr. D. Mackay;
in Stranraer: Miss A. Reid, Miss E. Ritchie;
at STUA: Miss B. Andrian, Dr. N. Dixon;
outside Scotland: Dr. T. Bayliss-Smith (Cambridge), Prof. J.M. Coles (Exeter), Dr. R. Gillespie (Dublin), Mr. D. Goodburn (Sittingbourne), Dr. A.R. Hands (Oxford), Dr. G.A. Lageard (Crewe), Miss C. Lavell (London), Prof. G. Lerche (Copenhagen), the late Mr. E. Macgillivray (Stockholm), Dr. J.P. Wild (Manchester), Mrs F. Wilkins (Kidderminster), Ms. B. Wolstrup (Copenhagen).

Last, but by no means least, the greatest possible thanks must go to my wife, Sue, and family, Adrian and Elinor, for tolerating my preoccupation with logboats over so many years.

As ever, all errors and omissions remain the responsibility of the author, who would, naturally, be grateful to hear of any new discoveries or suggestions for further research.

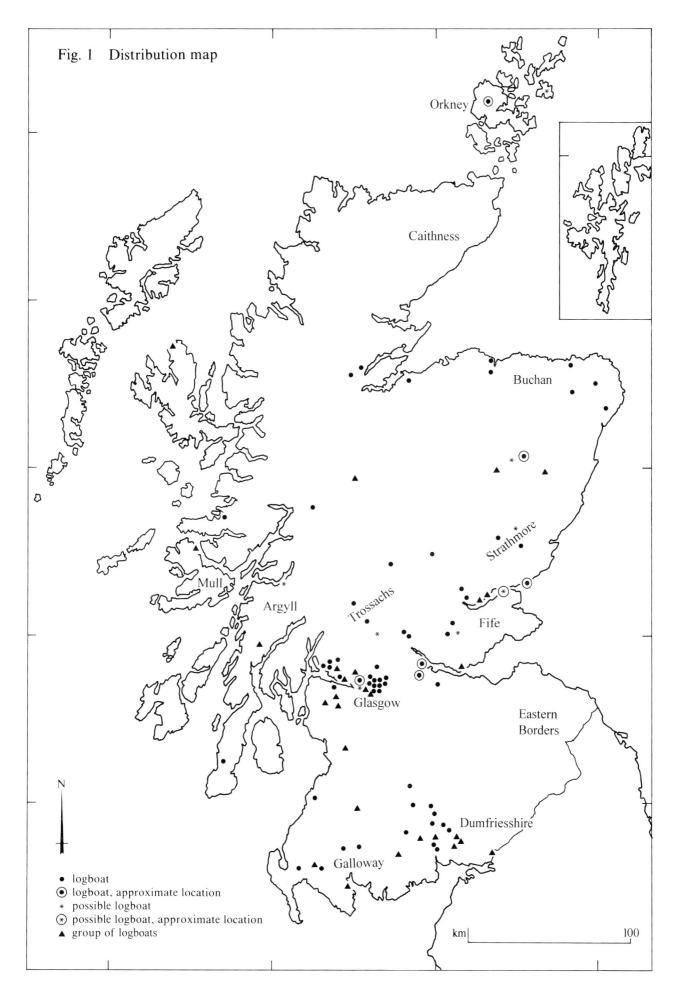

*Fig. 1 Distribution Map. Distribution of logboat discoveries, including possible examples, approximate locations, and groups of finds. Scale 1:212,000.*

# II Gazetteer of Logboat Discoveries

**1 Acharacle**
NM *c.* 6745 6822
NM66NE 1
Highland Region - Lochaber District
Argyll - Ardnamurchan ph.

Between 1895 and 1905 a logboat was discovered during drainage operations 'at the top end of the glebe, near the school'. The area indicated is on the W side of Acharacle village at an altitude of about 15m OD and is about 300m from the W end of Loch Shiel (at its present level).

No contemporary record of this discovery exists but a later account describes the boat as being 'narrow' and measuring between 12' (3.7m) and 14' (4.3m) in length. It was apparently sent to 'a Museum', but is now lost.

NMRS MS/47/1.


**Alloa** See **River Forth**


**Armannoch** See **Arnmannoch**


**2 Arnmannoch**
NX *c.* 888 752
NX87NE 10
Dumfries and Galloway Region - Stewartry District
Kirkcudbrightshire - Lochrutton ph.

The previous discovery of 'Part of a prehistoric canoe, dug out of a field' on the farm of Arnmannoch (formerly Armannoch) is mentioned in an account of 1901. The remains were possibly used as the cover of a well.

Arnmannoch farm is situated at an altitude of about 85m OD in an area where there is much drained land.

*TDGNHAS*, 2nd series, xvii (1900-5), 115.


**3 Auchlishie**
NO 3873 5788
NO35NE 5
Tayside Region - Angus District
Angus - Kirriemuir ph.

Between 1791 and 1820 what was possibly a logboat was found in a souterrain in the field known as Weems Park in Auchlishie. The discovery of 'querns' is also noted in the souterrain which is listed by Wainwright as his Kirriemuir II. It is situated 240m ENE of Kintyrie farmhouse and on the edge of a marked drop to the Quharity Burn in an area of rolling clayland and much drainage activity, at an altitude of about 150m OD.

It may be inferred from its being 'cut up for firewood' that the 'currach, or boat' contained a considerable quantity of timber in its construction and so was not of skin-based type. The narrow form of a logboat would presumably be more amenable to storage in a souterrain than that of a plank-built vessel.

*NSA*, xi (Forfar), 177, 178; Wainwright 1963, 191-2.


**Auchlossan Moss** See **Drumduan**


**4-5 Barhapple Loch 1-2, and paddle**
NX 2595 5915
NX25NE 2
Dumfries and Galloway Region - Wigtown District
Wigtownshire - Old Luce ph.

In 1878 drainage operations revealed a crannog in Barhapple Loch, which is situated in an area of rough pasture and extensive drainage at an altitude of about 85m OD. When it was subsequently excavated, the following objects (which are now lost) were found:

1. In 1880, 'two broad pieces of oak', each measuring about 4'6" (1.4m) in length, were found beneath a layer of stones. They were identified as possibly parts of a logboat. The published account is confused, and it is unclear whether they were built into the crannog, or were discovered about 150m to the NNE (at NX *c.* 2598 5929).

2. In 1884 a 'broken canoe paddle and half a canoe' were discovered during further excavation of the crannog.

Wilson 1882, 54-5; Munro 1885, 120.


## 6  Barnkirk
NX *c.* 39 66
NX36NE  14
Dumfries and Galloway Region - Wigtown District
Wigtownshire - Penninghame ph.

About 1814 a logboat was discovered during drainage or peat-digging operations in the Moss of Barnkirk which was formerly situated in the valley of the River Cree, at an altitude of between 30m and 50m OD and within the area of the former Forest of Cree. The boat was found at a depth of at least 6' (1.8m) and the discovery of what was apparently a ball of bog butter above it was probably coincidental.

The dugout interior of the boat measured 6'7" (2m) in length, 1'7" (0.5m) in breadth and 10" (250mm) in depth. The bow and stern were probably both 'a little rounded', and there were toolmarks in each of them. The boat was subsequently incorporated into a building on Barnkirk farm. It cannot now be located.

*NSA*, iv (Wigton), 178.


## 7  Barry Links
NO *c.* 53 32
NO53SW  39
Tayside Region - Angus or City of Dundee District
Angus - Barry or Monifieth ph.

About 1820 a logboat was revealed by the digging of a 'deep drain' across the links to the E of Monifieth. It was 'deeply imbedded' in a layer of buried peat at a point over a mile (1.6 km) from the shoreline.

*NSA*, xi (Forfar), 557; Malcolm 1910, 8.


**Bents**  See **Larg**

**Bishopbriggs**  See **Littlehill**


## 8  Black Loch
NS *c.* 797 107
NS71SE  34
Dumfries and Galloway Region - Nithsdale District
Dumfriesshire - Sanquhar ph.

About 1861 a logboat was discovered when the Black Loch was partially drained to recover the body of a suicide. This loch is situated about 1.6 km NE of Sanquhar at an altitude of 235m OD and has also been known as Sanquhar Loch or the Loch of Sanquhar. The crannog in the loch (at NS 7970 1068) has been approached by a causeway from the shore.

The logboat was found 'embedded in the mud, a few feet from the shore on the north side'. A 'long rounded pole' (possibly a punt pole) was found with it. The boat was of 'oak' and measured 3' (0.9m) in beam amidships and 1'10" (0.55m) at the bow; it was 16' (4.9m) long. There was a possible rowlock on one side, and in the bottom there were five thickness-gauge holes, each of them 'filled with a plug'. At the stern there were two 'well rounded holes' which Simpson suggests were used with the punt-pole, although the way in which he envisages its use is unclear.

The boat was coated with pitch to conserve it, but it had 'shrunk to very small dimensions' by 1891, and is now lost.

Jardine 1865, 4, 5; Simpson 1865, 170-1; Brown 1891, 32-3.


**Bog of Gight**  See **Gordon Castle**


## 9-10  Bowling 1-2, and paddle  (possible)
NS *c.* 44 73
NS47SW  12
Strathclyde Region - Dumbarton and Renfrew Districts
Dunbartonshire - Old Kilpatrick ph. and
Renfrewshire - Erskine ph.

In 1868 two logboats were lifted from the River Clyde during dredging operations 'in the...bend eastwards from Dunglass Castle'; they lay abreast about 2m apart with their prows to the SW. The 'club' that was found in one of them was said to be similar to those used locally for stunning salmon, but may have been a paddle. Neither of them has apparently been preserved.

1. The larger of the boats measured 23'6" (7.2m) in length and about 11' (3.4m) in 'mean girth', and was wrought from a bent oak log. On the 'bottom side' (presumably at the forefoot) there was a 'formidable projection' which extended forward some 2' (0.6m) and was pierced vertically, possibly to receive a painter or boat-rope.

2. The other boat measured 13' (4m) in length, 3' (0.9m) in breadth and 2' (0.6m) in depth and was 'carefully finished'; the sides were 'fitted for rowlocks' and there were two foot-rests near the stern. This was apparently a relatively beamy craft, having a slenderness index of about 4.3, and was possibly wrought from a squared whole section of a log.

*Geological Magazine*, vi (1869), 37; Buchanan 1883, 76-7; Bruce 1893, 19-20.


**Bowton**  See **Kinross**


**Buiston**  See **Buston**


11-12  **Buston 1-2 and oar**
NS 4155 4352
NS44SW 2
Strathclyde Region - Cunninghame District
Ayrshire - Dreghorn ph.

This crannog (which has also been known as Buiston, Biston, Mid Buiston and Swan Knowe) is situated within the area of the drained Loch Buston in the mid-Ayrshire clayland at an altitude of 90m OD. It is visible as a surface depression in pasture, and is seasonally flooded. It was discovered and heavily damaged during the reclamation of the former loch, when some 'thirteen cart-loads of timber' were removed from it and many of them were seen to be mortised. See also no. 13.

In 1880-1 Munro conducted excavations which initially comprised a long trench from NW to SE across the centre of the crannog; this may not have penetrated the substructure but revealed an extensive 'rubbish-heap' or midden on the SE. The 'central portion' of the crannog was then defined by annular excavation. The digging of sondages revealed 'layers of the stems of trees, chiefly birch' intermingled with brushwood, heather, moss, soil and large stones, beneath which there was a 'log pavement' of 'wooden beams like railway sleepers' retained by tenons. Beneath the margins of this deposit were identified three or four concentric circles of piles which were linked by both radial and annular mortice-jointed horizontal timbers. The uprights of a timber roundhouse with a central hearth, at least one other fireplace and flat stones 'covered' with slag were also identified, the last discovery being evidence for metal-smelting. Excavation along the supposed line of a causeway and at presumed location of a landing-place revealed an entrance-platform and a logboat.

The artefacts then recovered (NMS HV 1-201) include decorative objects of bronze and gold, bone combs, iron objects and the contemporary forgery of an Anglo-Saxon gold coin of 7th-century date, as well as pottery including false Samian, class E and probable Frankish wares. Numerous artifacts of worked stone and bone were found but there were few pieces of worked timber and few animal bones. Virtually all of these artifacts were found within (or derived from) the 'refuse-heap' that Munro identified on the SE perimeter of the crannog.

More recently, comprehensive excavation has been carried out (by Crone) as part of a regional programme of deterioration assessment (under the auspices of Historic Scotland). The following more complex sequence of structural phases was defined:
(I) Construction of the primary core by forming a mound of alternating layers of turves and brushwood over a primary layer of large boulders and massive oaken beams within a circle of oak stakes.
(II) Rebuilding of the central structure, with an extension to NW and the construction of an outer stockade comprising two concentric circles of squared stakes. The series of three superimposed floors and sub-rectangular hearths that is attributed to this phase forms the earliest evidenced period of occupation.
(III) Period of abandonment possibly precipitated by the slumping of the sub-structure. The crannog was extended to the NW and the extension levelled before the focus of occupation moved in that direction. A round house was built over a foundation of oak planks, fire-shattered stone and brushwood. The stone hearth at its centre was surrounded by floors of clay and brushwood; both floor and hearth were replaced at least four times. Also during this phase, a complex palisade of conjoined horizontal stakes was erected around the crannog; this structure was preserved in part by falling outwards on the SW side. The NW extension also gradually slumped outwards causing the collapse of the phase III house.
(IV) After this collapse, the crannog was again abandoned for a period before construction re-commenced in the NW quadrant. A massive framework of birch and oak logs was laid directly onto the lakebed sediments. The resulting log pavement may have formed a walkway between a defensive palisade and the walls of an inner building. The hollow into which had slumped the earlier structure became filled up with fragments of structural debris; most of the excavated artifacts were found in this area, which was presumably the source of the timber framework recognised by Munro.

The following radiocarbon determinations were obtained from samples obtained during the later excavations:

Charcoal or hearth ash from the uppermost of the group of
hearths, dating rebuilding of
central structure (phase II).
310 ± 50 ad  GU-2688

Pile or stake from timber
framework (phase IV).
520 ± 50 ad  GU-2636

Timber from outer ring of
palisade (phase IV).
370 ± 50 ad   GU-2637

Hearth ash from unstated context.
270 ± 50 ad   GU-3004

Brushwood from unstated context.
0 ± 50 ad   GU-3000

Successive excavations have yielded at least two logboats and an oar:

1. An extended logboat was found (by Munro) about 3m outside the structural timbers and adjacent to a possible landing-place at a depth of 6' (1.8m). The account of the discovery is unclear as to the orientation of the site; the published report places the logboat on the SW side of the crannog but the accompanying site-plan depicts it in the SSE. The boat was probably taken to the Dick Institute, Kilmarnock where a logboat from Buston is recorded as having been destroyed in the fire that severely damaged the building on the night of 26 November 1909.

The unusual construction, evidence of propulsion and wide variety of repairs make this vessel of exceptional interest. It was of 'oak' and measured 22' (6.7m) in length over all, and 19'6" (6m) internally; the internal depth measured centrally was 1'10" (0.6m). The internal breadth was 3'6" (1.1m) at the stern, 4' (1.2m) amidships and 2'10" (0.9m) at a point 'near the stem'.

Munro's published drawing depicts a boat of flared rectangular section which was apparently worked from a half-sectioned log of about 1.2m diameter. The boat was narrow, having a slenderness coefficient of 5.5, and (on the assumption that the angle of flare was about 20° as the drawing indicates) the seaworthiness index was of the order of 2. The conversion coefficients were 89% from the whole log and 78% from the half log, which figures fall around the mid-point of the anticipated range. The displacement in the standard condition was probably around 2.6 cubic metres. The drawing depicts the boat from above the stern so that the form of the bow is difficult to ascertain, but the McGrail morphology code was probably 44a3:313:223, making it a variant example of the dissimilar-ended form.

The bow of the boat was rounded with a projecting stem which was pierced horizontally by a large hole, and the stern was formed by a 'strongly constructed' semi-circular transom which measured 3'6" (1.1m) across the beam, about 1'4½" (0.4m) in depth and 3½" (90mm) in thickness. This was set into a groove and held into position by slender bars of wood mortised into the sides in front and behind. About 3' (0.9m) from the bow, each side was pierced by an elongated hole 'near the rim', which was possibly a hand-grip.

About 1'3" (0.4m) forward of the transom (at about 93% of the length) there was a false rib left in the solid across the sides and bottom, possibly for additional strength. There were also 'one or two round holes', probably for thickness gauges, in the floor.

A 'sort of gunwale' had been pegged externally onto the sides 'from within a few feet of the stem till it projected a little beyond the stern'. This feature is noted in the published account but omitted from the drawing. In two places 'equidistant from the ends, and about 4 feet (1.2m) apart' there were 'short pieces of wood fastened to it by vertical pins, as if intended for the use of oars' set into the tops of the sides. There were also holes set 'along its upper edge, as if for thole pins'.

Extensive measures had been taken to strengthen the boat, possibly in consequence of its poor construction. Along the greater part of one side thin softwood boards had been inserted within the oak skin. Both the numerous splits in the oak and the space between the skin and the lining had been stuffed with moss. Numerous 'well-shaped ribs' had been inserted at irregular intervals, possibly with a concentration towards the stern. Two of them were of oak and the rest of greatly-decayed birch or similar softwood; all of them were retained by closely-spaced wooden 'pins' or treenails. Two 'nicely-fitting' repair-patches had been inserted into the sides; the larger of them measured 2'3" (0.7m) in length and 10" (255mm) in breadth, and was held in position by two ribs.

In the brushwood layer above the logboat there was found an oak 'beam' of rectangular section measuring 5" (125mm) by 3½" (90mm) and which broke upon extraction. At intervals of 1' 10" (0.6m) along the narrow side of this timber there were three round holes containing the remains of broken pins. This was possibly a further section of the side of the vessel, or of the washstrake.

The discovery 'on the crannog' of what was probably an oar, serves to support the suggestion that the boat was so powered. Its form was not recorded but the blade was 9" (175mm) broad and 1¼" (30mm) thick and the handle measured 5" (125mm) in circumference.

Munro 1882b, 201, 203, 206-10, 270-1; NMAS 1892, 254-6; *Kilmarnock Standard*, 9 November 1911; Morrison [1975]; Scott 1976, 44; McGrail 1978, i, 37, 53, 59, 64-6, 68, 76; McGrail 1987a, 73, 78, 82.

2. This logboat was found in 1992 (by Crone) held in place on the NW side of the crannog, where it was transfixed by a stake associated with the construction of the final (phase IV) palisade.

Some 6.5m of the length of the boat was exposed (and subsequently re-buried) beneath about 2m of peat and clay; the excavator has tentatively suggested that the length overall was no more than about 9m. The exposed stern has suffered from post-depositional flattening and cracking but the original rounded profile (where exposed near the section) apparently survives to the full height of the sides and measures 0.7m in beam by 0.65m in depth, both measurements being apparently taken internally.

## II Gazetteer of Logboat Discoveries

*Pl. 1. Cambuskenneth (no. 14). Apparently-innacurate but artistically satisfying lithograph impression of the scene at the time of the discovery. Reproduced, by kind permission of the publishers, from the* Illustrated London News *of 6 June 1874. (NMRS DC 11554/p).*

The interior of the vessel is said to be 'beautifully worked with adze grooves, 9cm wide, running across the hull'. The transom has been lost but its groove remains clearly defined. Five roughly-rectangular projections along the sheerline may have been intended to retain thwarts, while 'a number of shallow circular impressions' located lower down the sides may indicate the former location of struts.

On the basis of the available evidence, the McGrail morphology code of the boat is 44a2:2x2:xxx, making it a variant of the dissimilar-ended form.

Munro 1882, 190-239, 245; NMAS 1892, 254-6; Scott 1976, 37, 38, 44; Morrison 1985, 36, 43, 44; *DES*, (1989), 59-60; Historic Buildings and Monuments, Scottish Development Department, news release no. 51/90, dated 9 August 1990; *DES*, (1990), 36; Crone 1991; Barber and Crone 1993, 522, 527, 528-30; Earwood 1993a, 89, 268; *pers. comm.* Dr. A. Crone.

### 13 Buston 3
NS *c.* 415 435
NS44SW 34
Strathclyde Region - Cunninghame District
Ayrshire - Dreghorn ph.

A further logboat is said to have been discovered at an unknown date on Lochside farm and near nos. 11 and 12. Its association (if any) with the crannog is unclear.

It was at first taken to the Dick Institute, Kilmarnock, but its present whereabouts are unknown and it may have been transferred to the Hunterian Museum, University of Glasgow. See also no. 149.

*Pers. comm.* Mr. J. Hunter.

### Cadder Moss  See Littlehill

### 14 Cambuskenneth
NS 805 939
NS89SW 28
Central Region - Stirling District
Stirlingshire - Stirling or St Ninians ph.

In May 1874 a logboat was discovered (pl. 1) in a mean-

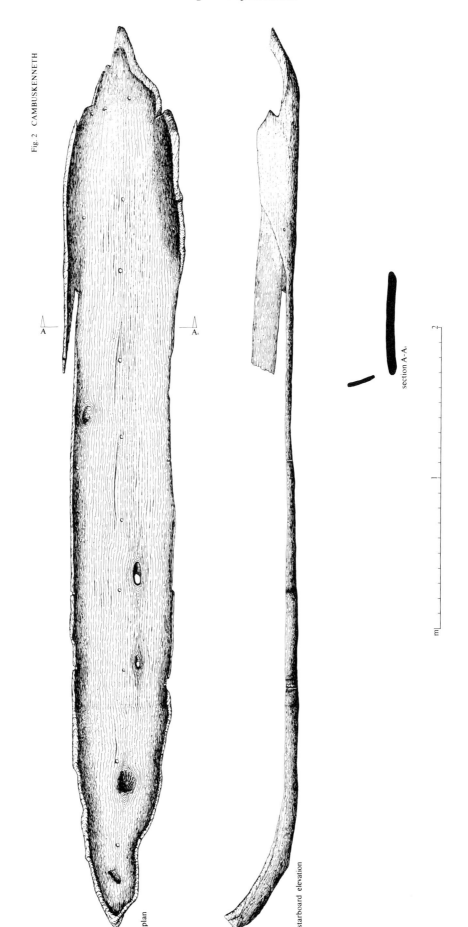

Fig. 2 Cambuskenneth (no. 14). Logboat. Plan, starboard elevation and section at 33% of length from bow. Scale 1:25

der of the River Forth to the E of Stirling during a period of low water level and at a point 'about fifty yards (46m) below Cambuskenneth Abbey Ferry...in a direct line between the Abbey Mill and Hood farm-steading'. The finder was William Johnston, a local publican and a 'great searcher of antiquities' who subsequently complained bitterly in the local newspaper about 'the clumsy manner in which the canoe has been excavated from the river's bed' with the result that a 'piece of her side, about five feet (1.5m) long, has been recklessly broken off'. The discovery was variously identified as that of a 'Caledonian' vessel and as that of a ferry boat which capsized tragically in 1529; some of the local people, however, claimed to remember the manufacture of its nails.

The vessel was found to measure about 20' (6.1m) in length although incomplete, the 'prow' having broken away. The beam was about 3' (0.9m) and a broken piece of the side was found inside it. The thickness of the sides and bottom was noted as between 1" and 2" (25mm and 50mm). Several holes were noted which had 'evidently been carefully plugged up' and the timber was identified as 'oak'. Evidence for damage in use was found; one of the sides had been patched near one of the ends with a piece of wood measuring about 8" (220mm) square which had been fastened in place with 'broad headed nails made of malleable iron'. The boat was re-recorded by Feachem about 1958.

The boat (fig. 2) is displayed in Cambuskenneth Abbey, which is a monument in the guardianship of Historic Scotland; the bottom of the boat could not be inspected at the date of the visit. The pattern of radial splitting on one end suggests that it has been worked from a split half-section of log which was largely free of knots. Splitting has occurred along the junction of the sides and the bottom while warping and shrinkage have caused considerable distortion and large sections of the sides have broken away, including part of the portion that Feachem depicted as partly detached (and which is not drawn for publication). The patched section has not survived. There is extensive charring of unknown origin along the exterior of the higher surviving sides and within the interior there are isolated toolmarks which were probably made during recovery operations.

The boat measures 6.1m in length over all and up to 0.78m in beam, and the sides are nowhere more than 90mm high above the floor. Both of the ends are inclined upwards and the more rounded of them is identified by Feachem as the stern, but (on the assumption that the lower part of the trunk was used to form the stern) the direction of slant of the knot-holes suggests the converse to be the case. The flat bottom measures between 50mm and 75mm in thickness and the surviving portions of the sides considerably less.

The timber is extensively pierced by both knot-holes and holes for thickness-gauges. There are nineteen of the latter, disposed in three lines, one of them (bored vertically) down the centre-line and the other two (comprising holes bored down and out towards the sides) in the near right-angles between the sides and the floor. Each hole measures about 30mm in diameter. The possibility that at least some of these holes were intended to retain fitted ribs cannot be ruled out.

The McGrail morphology code of the boat is 332:1x1:332, making it of variant canoe (or possibly punt/barge) form. On the basis of the evidence available, the slenderness index is 7.8.

This boat has been dated by radiocarbon assay to 915 ± 45 ad (GrN-19281), which date may be calibrated to about 996 cal AD. *December 1987.*

*Stirling Observer*, 14, 21 and 28 May 1874; *Illustrated London News*, 6 June 1874, 537-8, 544; Feachem 1959; McGrail 1987a, 65; *pers. comm.* Dr. J.N. Lanting.


## 15 Carlingwark Loch
NX *c.* 76 61
NX76SE 8
Dumfries and Galloway Region - Stewartry District
Kirkcudbrightshire - Kelton ph.

The *Statistical Account* notes the discovery at 'several places' in Carlingwark Loch of 'canoes' which had been 'hollowed...with fire'. The loch is situated in pastoral drumlin country at an altitude of about 45m OD and the boats were probably found either in 1765 when the size of the loch was greatly reduced by the construction of a drainage-canal, or in the following years when much of the lake-bed sediment was taken for agricultural marl. Their fate is not recorded.

Other antiquities found in the loch include Iron Age and Romano-British metalwork and at least one crannog, but there is no evidence to support the 'Celtic' attribution of the boats that Affleck proposes.

In view of the experimental evidence for the ineffectiveness of fire in working large oak timbers (*pers. comm.* Damian Goodburn) these boats may have worked from softwood or (more probably) fire-working may have been suggested on the basis of the mis-interpretation of incipient rot.

*Stat. Acct.*, viii (1793), 306; Wilson 1851, 31; Affleck 1912, 237; Reid 1944, 50; MacGregor 1976, ii, nos. 268, 287, 309.


## 16 Carn an Roin
NN 0648 2273
NN02SE 15
Strathclyde Region - Argyll and Bute District
Argyll - Glenorchy and Inishail ph.

In 1972 underwater archaeological survey revealed what was possibly the bottom of a logboat 'embedded in the surface stones' on the NE side of a crannog and about 5m from the waterline. It lay radially, with the visible pointed end towards the centre of the crannog, and may have formed an outer structural timber. The remains were left *in situ* and not recorded in detail.

The crannog is situated 480m E of Achnacarron farmsteading, near the NW shore of Loch Awe, a major freshwater loch which is situated in a glaciated valley at an altitude of 36m OD.

*DES*, (1972), 11-12; Hardy *et al*. [1972], 10; McArdle, CM and TD (1973), 6; RCAHMS 1975, 94, no. 198 (5); *pers. comm.* Dr. D. McArdle.


**Carron Valley** See **River Carron**


17 **Carse Loch**
NX *c*. 919 846
NX98SW 68
Dumfries and Galloway Region - Nithsdale District
Dumfriesshire - Dunscore ph.

In 1879 Munro examined a 'canoe' in Dr. Grierson's Museum at Thornhill. It had been revealed by drainage operations about 60 yards (55m) from the crannog or 'island' in Friar's Carse and was found 'deeply imbedded' in the mud at a depth of about 4' (1.2m).

Friar's Carse farmsteading is situated at NX 9254 8499 and a crannog is noted in Carse Loch at NX 9189 8465. The local topography is cultivated clayland at an altitude of about 30m OD.

On discovery, the boat probably measured about 22' (6.7m) in length and 2'10" (0.9m) in beam, indicating a slenderness coefficient of about 7.8. The bow had been formed from the root end of the tree and the stern was closed by a transom, making it of dissimilar-ended form.

No subsequent record of this boat can be found.

Munro 1882c, 73, 76-7; Simpson 1895, 29.


**Carse of Falkirk** See **River Carron**


**Castle Loch** See **Closeburn**


18 **Castle Semple Loch**
NS *c*. 36 59
NS35NE 6
Strathclyde Region - Renfrew District
Renfrewshire - Lochwinnoch ph.

The discovery of large numbers of possible logboats has been reported from Castle Semple Loch (or Lochwinnoch) which is situated in an industrialised valley at the head of the Black Cart Water, and at an altitude of about 28m OD. None of these remains survive.

Some of the timbers found were probably those of a crannog, but the *Statistical Account* notes several that were found about 5' (1.5m) 'below the surface' and were compared to the canoes of the North American Indians. These were most probably logboats.

*Stat. Acct.*, xv (1795), 68; *NSA*, vii (Renfrew), 97; Stuart 1866a, 159, 174; Love 1866, 289.


19 **Castlemilk**
NS *c*. 60 59
NS65NW 4
Strathclyde Region - City of Glasgow District
Lanarkshire - Carmunnock ph.

About 1831 what was possibly a logboat was discovered during ditch-digging 'on the march between Cathkin and Castlemilk' in a hilly area which varies in altitude between 50 and 100m OD. The 'bottom of a boat' measured 10' (3.1m) in length by 2' (0.6m) in breadth and was 'all of black oak'; 'strong wooden nails' (which were presumably treenails) were also found.

The discovery is discussed by both Wilson and Stuart. The former describes it as 'of more artificial construction' than the Glasgow logboats and sees it as being 'secured by large wooden pins'. Stuart, however, classifies it as a 'canoe'. It is now lost.

*NSA*, vi (Lanark), 601; Wilson 1851, 37; Stuart 1866a, 149, 174.


20 **Catherinefield**
NY *c*. 0013 8011
NY08SW 18
Dumfries and Galloway Region - Nithsdale District
Dumfriesshire - Dumfries ph.

In 1973 a logboat was revealed during the mechanical re-cutting of the Old Course of the Lochar Water across Lochar Moss in middle Nithsdale at an altitude of about 15m OD and in an area which had previously been extensively drained. Part of the boat broke away during digging operations and was left in the open air for some time before its significance was recognised by a visiting geologist. It was initially removed to a storage tank in Dumfries and is currently (October 1994) undergoing conservation in the laboratories of the National Museums of Scotland prior to return to Dumfries Museum for display under accession number DUMFM 1974.182. The remainder of the boat is presumably still *in situ*.

The recovered section measures up to 2.24m in length and between 0.75 and 0.81m in breadth; it has been hol-

## II Gazetteer of Logboat Discoveries

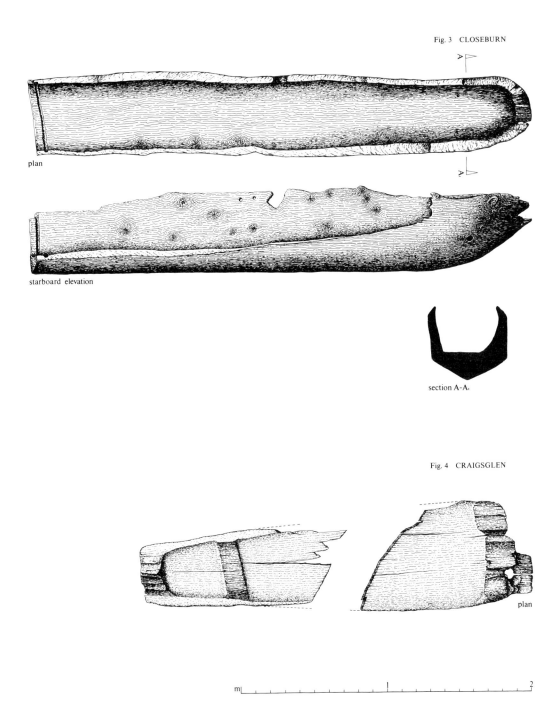

*Fig. 3. Closeburn (no. 21). Logboat. Plan, starboard elevation and section at 12% of length from bow. Scale 1:25.*

*Fig. 4. Craigsglen (no. 23). Logboat. Plan view of surviving remains. Scale 1:25.*

lowed from a split (and presumably half-sectioned) trunk of oak which probably measured about 1m in diameter. The starboard side has been considerably damaged and the floor less so, but the port side may survive in part to its full height, which suggests an internal depth of 0.23m. The sides are slightly flared and measure between 30mm (at the top) and 80mm (at the bottom) in thickness. The floor is about 0.18m thick and the stern has been formed by a flat-topped block measuring about 0.32m in length. In the absence of a large part of the boat the slenderness coefficient can only be said to have been at least 2.9. On the assumption that the angle of the flare was 10°, the seaworthiness ratio appears to be about 0.06. The McGrail morphology code may be cited as 533:xx3:xxx but the form cannot be ascertained on the basis of the incomplete remains.

The boat has been radiocarbon-dated to 1804 bc ± 125 (SRR-326), which determination may be calibrated to about 2143 or 2183 cal BC.

Jardine and Masters 1977; McGrail 1987a, 86; *Glasgow Herald*, 24 December 1990; *Scotsman*, 31 December 1990; *Scotsman*, 14 January 1991; *pers. comm.* Mr. D. Lockwood.


**Causewayhead** See **River Forth**


**21 Closeburn**
NX *c.* 906 921
NX99SW 4
Dumfries and Galloway Region - Nithsdale District
Dumfriesshire - Closeburn ph.

On 5 April 1859 a logboat was retrieved during the drainage of the NE part of the former Castle Loch at Closeburn in upper Nithsdale, at an altitude of about 60m OD. A bronze tripod and a paddle had previously been found in the loch and on the E shore there is a fourteenth-century tower-house. A second paddle (no. A29) was found three years later at Kirkbog in Closeburn parish.

The logboat was found in its conventional attitude in the peat at a depth of about 0.9m and was aligned E-W. Adam noted that it was of 'oak' had a groove-mounted transom. It measured about 12' (3.6m) in length and 2' (0.6m) in beam.

The logboat (fig. 3) is housed in the reserve collection of the Royal Museum of Scotland under accession number NMS IN 2. It has twisted noticeably (presumably in drying), and the transom and most of the starboard side are missing. It measures (after shrinkage) 3.46m in length, up to 0.54m in beam, and about 0.5m in depth from what may be full height of the side. The port side measures about 40mm in thickness near the base and 20mm higher up. The floor of the boat is flat throughout its length and the sides are vertical, making the internal cross-section roughly square. The bottom measures about 100mm in thickness at the stern but is considerably thicker along the centreline forward, where the underside has a noticeable V-form.

On the basis of these measurements, the boat has a slenderness coefficient of 6.4 and a beam/draught coefficient of about 1.1. The displacement under standard conditions was about 0.56 cubic metres.

The transom-groove measures between 25mm and 30mm in breadth and between 20mm and 25mm in depth; it has probably been cut with a metal chisel. The bow is rounded both internally and externally. Around it there is evidence of charring and a rough area of heartwood, and beneath it there are broad toolmarks which have probably been made with an adze. Six probable thickness-gauge holes (four of them displaced towards the stern) pierce the floor, and in the sides near the bow there are two further holes of irregular form and undetermined nature.

The timber is noticeably smooth, fine-grained and free from splitting, but has numerous knots, the direction of which indicates that the bow is at what was formerly the lower part of the tree. The vessel has been constructed eccentrically from a complete parent log which measured at least 0.55m in diameter. About 94% of the timber has been removed to leave a useable volume of approximately 0.86 cubic metres. The McGrail morphology code is 411:111:321 and the form is a variant of the dissimilar-ended type.

This boat has been radiocarbon-dated to 1140 ± 50 ad (GrN-19279), which determination may be calibrated to about 1235 cal AD. *May 1978.*

*TDGNHAS*, 1st series, (1862-9), 7, 34; *PSAS*, vi (1864-6), 435; Adam 1866; Munro 1882b, 245, 280; *pers. comm.* Dr. J.N. Lanting.


**22 Clune Hill, Lochore**
NT *c.* 16 95
NT19NE 31
Fife Region - Dunfermline District
Fife - Ballingry ph.

In 1926 the 'remains of a dug-out canoe' were found 'near the Clune Hill' and within the former area of Loch Ore, where the remains of a possible crannog or lake-dwelling have also been noted. The vessel was apparently 'identified by a local teacher' (but not described in detail) before being sawn up for miners' firewood.

The area has been heavily disturbed by industrial activity and the loch that is now to be seen has been formed by mineral extraction rather than natural processes. Clune Hill cannot be specifically identified, but Clune Craig is on the S side of the present loch, at NT 16 95.

Henderson 1990; *DES*, (1993), 27.

**Craigs Moss** See **Lochar Moss**

**23 Craigsglen**
NJ *c.* 78 56
NJ75NE 5
Grampian Region - Banff and Buchan District
Aberdeenshire - King Edward ph.

In about 1893 a logboat was discovered during drainage operations in the Glen of Craigston or Craigsglen, a steep-sided valley which cuts into the plateau of NW Aberdeenshire at an altitude of about 100m OD. The lack of navigable water in the area is probably the result of post-medieval drainage.

The logboat was not at first recognised as such and it was cut into three parts for ease of removal; the centre section was destroyed but the two ends were taken to Craigston Castle. In 1951 the remains were passed to Aberdeen University Anthropological Museum where they are in store under accession number AUAM 576-1.

Both the surviving portions (fig. 4) display evidence of warping and splitting, most noticeably at the ends. Neither of them has thickness-gauge holes or obvious toolmarks. The bottom of the boat measures about 40mm in thickness and curves upwards into the sides.

The portion that is in better condition measures 1.41m in length by up to 0.51m transversely, and was probably the forepart. The presumed bow is near-rectangular on plan, and the timber forms a flat shelf measuring 110mm in breadth and 110mm in height internally. Beneath the boat there is what may be a broken hole measuring about 80mm in diameter, and in the interior, about 0.4m from the end, there is a false rib left in the solid; this feature measures about 110mm in breadth across the flat top and 60mm in height.

The remains of what was probably the after section measure 1.19m in length by up to 0.7m transversely. The solid end is about 110mm high internally, and its flat top measures about 210mm across. An external extension of unknown purpose projects off-centre beyond the end; this measures about 0.15m in length and through the junction of this extension with the solid end there is a hole of unknown function measuring about 80mm in diameter.

Assuming the narrower end to be the bow, the McGrail morphology code of this vessel is 112:2x2:112, making it a variant of the canoe form. *September 1987.*

Godsman [1952], 286-7.

**24 Croft-na-Caber**
NN 769 448
NN74SE 30
Tayside Region - Perth and Kinross District
Perthshire - Kenmore ph.

In 1994 the remains of a logboat were discovered during the experimental construction of a replica crannog for public exhibition (Andrian 1995) in shallow water off the Croft-na-Caber watersport activities centre near the NE end of Loch Tay. It is impossible to exclude the possibility that this is a re-discovery of the possible example noted at Portbane (no. 144).

Loch Tay is situated in a deep glaciated valley in the Trossachs at an altitude of 107m OD. Within the extensive area of its waters there are at least seventeen crannogs (one of which, at Oakbank, has been excavated) and a medieval priory-island, but none of these monuments is in the vicinity of the logboat. The published attribution of this boat to Oakbank crannog (for which see no. A56) is erroneous.

The vessel has been recorded (and remains) *in situ*. It is of oak, in two portions, and measures between 8m and 10m in length by 0.85m in breadth. Caulking material and toolmarks have been recognised within the transom-groove.

*Scotsman*, 20 July 1994; *Perthshire Advertiser*, 2 September 1994; *STUA News*, no. 4.1 (summer/autumn 1994), 3; *IJNA*, 24 (1995), 219; *pers. comm.* Dr. T.N. Dixon.

**Cruive Bank** See **Lindores**

**25 Dalmarnock**
NN *c.* 998 458
NN94NE 10
Tayside Region - Perth and Kinross District
Perthshire - Little Dunkeld ph.

On 8 May 1975 a logboat was discovered by contractors engaged in the construction of the new route of the A9 trunk road across the haughland of the middle course of the River Tay to the S of Dalguise village and at an altitude of about 55m OD. The boat lay upside-down just below the surface of the gravel and was removed to a lagoon for recording. It has since been placed in a nearby loch for storage.

Abrasion has reduced the boat (redrawn from the excavators' drawing as fig. 5) to its nearly-flat bottom, which measures 4.58m in length, 0.82 in beam and between 80mm and 100mm in thickness. Both bow and stern are rounded and there is part of a possible projecting sternpost. Near the end which was probably the stern and also 2.14m from it there are two possible thickness-gauge holes measuring about 25mm in diameter; when discovered these were 'closed by reddish chert pebbles rammed in'. The lack of biconicality of these holes has been used to infer the use of a metal drill and consequently a medieval or later date.

When visited by Niall Gregory on 6th August 1992,

the boat was lying perpendicular to the shore with about half of its height having been exposed above the surface by a fall in the water level of the loch, the narrower end (and presumed bow) being under water. The surface texture of the upper part was becoming flaky, but little splitting and no warping were evident. The rounded ends were identifiable and the upturned stern in section, while the timber was seen to have be pierced by several small knots, a probable thickness-gauge hole and the locations of two archaeological samples. In its current state, the vessel measures 4.47m in length over all and up to 0.79m in beam (measured at a point 1.3m from the presumed stern). The thickness of the floor varies between 45 and 50mm.

A visit by Mark Hall and others on 28 July 1995, revealed the boat to have been almost entirely beached, the upper surface having become flaky and the exposed end (the presumed stern) being covered with moss and having incipient cracks. Before the boat was re-submerged, it was re-measured and the length recorded as 4.54m, while the thickness of the timber was variously recorded at between 50 and 80mm. The width was measured at the stern, amidships and at the bow, to record figures of 0.72m, 0.77m and 0.4m respectively.

On the evidence of the incomplete remains, the slenderness index was of the order of 5.5 and the form has possibly been that of a variant canoe. The McGrail morphology code is apparently 222:1x2:222, making it a possible variant example of the canoe form.

*DES*, (1976), 49-50; *International Journal of Nautical Archaeology*, 7.1 (1978), 86-7; NMRS MS/736/4-5, PTD/298/1-2, PTR/35/1 and A41747-9.


26 **Dalmuir**
NS *c*. 474 709
NS47SE 61
Strathclyde Region - Clydebank District
Dunbartonshire - Old Kilpatrick ph.

In June 1903 a logboat was discovered during the construction of the Sewage Purification Works at Dalmuir 'about 50 yards' (46m) N of the Clyde Trust Works. It was 'embedded in the sand near the present river margin' at a height of 2' (0.6m) above the low water mark, and in an area which had possibly been previously disturbed during the construction of an artificial river bank. The vessel was apparently brought to land, but its subsequent fate is unknown.

The discovery is described at some length in an unattributed (but apparently authoritative) newspaper account. The boat measured 'a little over' 15' (4.6m) in length, by 'about' 3' (0.9m) in 'breadth' but the 'top part' was destroyed before it was recognised. It was similar in form to the previous Clyde discoveries (presumably nos. 9-10, 35, 39-42, 46 and 136) and was said to have been wrought from 'a single oak tree, skilfully hollowed out by metal tools'. At a point 4" (100mm) from the stern (and about 98% of the length from the bow) there was a transom-groove which had been 'neatly and uniformly cut out to a depth and breadth of one inch (25mm)'. There were said to be numerous 'holes and wooden pins' of a high standard of workmanship, but the nature, locations and functions of these were not noted in detail. On the evidence of this account the slenderness index was of the order of 5.1 and the form of the boat was dissimilar-ended.

Four 'oblong pieces of wood' were found 'at the bottom of the canoe', where they were fastened with wooden pegs when first discovered; at least one of them was curved to fit the bottom of the boat. These were identified as footrests, but were more probably fitted ribs.

*Glasgow Herald*, 21 November 1903.


27 **Dernaglar Loch**
NX *c*. 263 581
NX25NE 6
Dumfries and Galloway Region - Wigtown District
Wigtownshire - Old Luce ph.

In 1885 Munro examined a 'single-tree dug-out' which had earlier been discovered during drainage operations around the margins of Loch Dernaglar (or Dernaglaur). This loch is situated in rough pasture among drumlins and at an altitude of about 90m OD. A crannog was revealed at the same time but there is no evidence that the logboat was associated with it, and the relative locations of the two discoveries were not noted.

The logboat was 11' (3.3m) long, 2'7" (0.8m) 'wide' and 11½" (290mm) 'deep'. A transom-groove was situated 9" (230mm) from the end and at 93% of the length of the boat. The interior was divided into 'compartments' by what were probably four fitted ribs, at least one of which was retained by a treenail.

It appears unlikely that the full height of the sides was discovered, but the slenderness coefficient appears to have been about 4.25.

Munro 1885, 121; Wilson 1885, 72.


28 **Dingwall**
NH *c*. 55 57
NH55NE 3
Highland Region - Ross and Cromarty District
Ross and Cromarty - Dingwall or Urquhart and Logie Wester ph.

In 1874 erosion revealed a logboat in the bank of the River Conon 'opposite Dingwall' and at a point which was possibly near the upper limit of tidal flow. It lay in silt at a depth of about 2.5m below the surface of a gravel bank.

The boat was of 'oak' and measured 16' 3" (5.0m) in

II *Gazetteer of Logboat Discoveries* 23

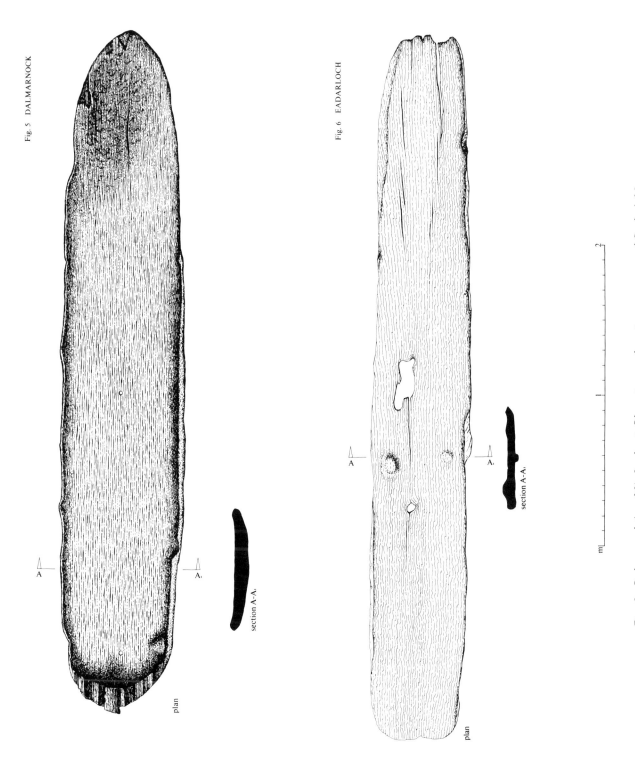

*Fig. 5. Dalmarnock (no. 25). Logboat. Plan view, and section near one end Scale 1:25*
*Fig. 6. Eadarloch (no. 36). Pos*

length over all, 2' (0.6m) in beam 'at the bow' and 2' (0.6m) in 'depth of the side'. The figure of 3' (0.9m) that is cited for the beam at the 'stem' probably refers to the stern. On the basis of the recorded measurements and assuming the full height of the sides to have been found, the slenderness coefficient of this boat was 5.4 and the beam/draught coefficient was 1.5. The displacement under standard conditions was about 1.6 cubic metres. These figures indicate that this was a medium-sized boat of non-specialist type worked from a whole-sectioned log.

The boat was subsequently donated to the museum of the Society of Antiquaries of Scotland (which was later incorporated into the collections of the Royal Museum of Scotland). It cannot now be identified in the collections.

*PSAS*, xvi (1881-2), 11.

### 29-33 Dowalton Loch 1-5
Dumfries and Galloway Region - Wigtown District
Wigtownshire - Sorbie ph.

In 1863-4 Lord Lovaine (later Earl Percy) and others investigated the recently-drained bed of Dowalton Loch, which occupied an area of about 4 square kilometres at an altitude of about 45m OD in the drumlin country of the Wigtownshire machars. Five logboats, at least eight crannogs, four bronze vessels and numerous other artifacts were revealed.

Logboats (1) to (3), and possibly also numbers (4) and (5), were donated to the then museum of the Society of Antiquaries of Scotland but cannot now be identified among the collections of the present Royal Museum of Scotland.

1. NX 4061 4681   NX44NW 2

This logboat was discovered in 1863 during Lovaine's excavation of the crannog that is situated 760m SW of Stonehouse farmsteading; it was built into the substructure on the NE side 'with hurdles and planks above it' and was 'very complete, and in good order'.

The boat (pl. 2, left) was apparently side-extended and measured 21' (6.4m) in length and 3'10" (1.2m) across the stern, which was of unusual construction. A plank had been inserted into a groove on each side and a backboard pegged in place above it; the depth at the stern was noted as 1'5" (0.4m) or 1'8" (0.5m) to the top of the backboard. Other unusual features included the two possible tholepins wedged into holes on each side and the 'plank or washboard' which ran round the boat and was pegged into the solid wood. The published description mentions one 'thwart of fir or willow' but one of the unpublished drawings depicts a second. The drawings depict a craft with a bow of rounded point shape and mid-section curved in all three planes. The significance of the horizontal lines that are depicted along the sides of the boat remains unclear.

Among the manuscripts of the Society of Antiquaries of Scotland there are three annotated sketches which depict this vessel in greater detail; two of them (one a 'restored' view) are drawn from the starboard quarter, while the third (which is also 'restored') is a plan view. The bow is depicted as of rounded point form and the detailed shading appears to indicate what may be two false ribs left in the solid at about 53% and 67% of the length from the bow; each of the single holes set into the upper surfaces (where they reach the full height of the sides) is shown as retaining a tholepin. The three washboards (one along each side and one across the stern) are depicted as projecting at an angle nearer to the horizontal than to the vertical, while the thwarts are nailed in place above them, the after one (at least) being bent up at the sides. The artist has labelled a 'piece of wood inserted with two pins' on the port bow, and this may be a repair-feature of some type.

The 'as discovered' quarter-view of the vessel shows the line along the starboard side to represent a pronounced longitudinal split, while the apparently-conjectural quarter view depicts the stern as having what is apparently a detached transom, restores the starboard washboard and ignores the longitudinal split in the same side.

Calculation on the basis of the recorded measurements indicates that this was a small but beamy craft which was probably 'volume dominated' and so well-suited to convey light but bulky cargoes. The displacement was about 1.3 cubic metres and the slenderness and beam/draught coefficients were 3.1 and 2.3 respectively. The McGrail morphology code was apparently 14b4:332:322 and the form was, in consequence, dissimilar-ended.

Stuart 1866a, 118; Munro 1885, 87; McGrail 1987a, 73, 83; Earwood 1993a, 273; NMRS WGD/49/1; SAS MS/198.

2. NX 4076 4694   NX44NW 3

Lovaine describes the discovery of a 'canoe of oak...surrounded by piles' on the N side of the crannog that was formerly situated 550m SW of Stonehouse farmsteading. Excavation revealed that the logboat lay within what was possibly a dock or similar structure.

The boat itself measured 24' (7.3m) in length, 4'2" (1.3m) in beam amidships and 7" (178mm) in depth. The floor of the boat was 2" (51mm) thick. On the self-evident basis that the sides were found incomplete, the only possible calculation is that of the slenderness coefficient, which falls within the mid-range of the sample at 5.76.

A further (undated) drawing in the manuscript collection of the Society of Antiquaries of Scotland is untitled but appears (on the basis of the annotated dimensions) to depict this vessel under the heading of 'No. 1. Larger Canoe'. The vessel is depicted in plan, starboard elevation and cross-section and is shown as having been reduced

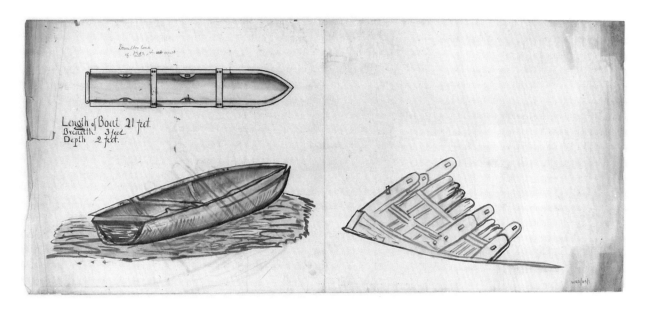

*Pl. 2. Dowalton Loch 1 (no. 29).*
*Left hand: plan and quarter-view of the logboat, probably drawn by Robertson or Percy at the time of the discovery, but not included in the published account. Right hand: evidently depicts the 'heavy slabs of oak...laid upon one another in a sloping direction, bolted together by stakes inserted in mortises...and connected by square pieces of timber' that secured the upper part of the south side of the crannog (Munro 1885, 86). Reproduced from a drawing in the NMRS (WGD/49/1/2; SAS439) by kind permission of the Royal Commission on the Ancient and Historical Monuments of Scotland.*

to a nearly-flat plank, with only a slight rise at the bow; at the stern there is an enigmatic (and unnoted) downwards slope which may represent a split originating from a transom-groove. The bow has evidently been of pointed or rounded point form, and the bottom rises steeply into the sides while the inner surface of the vessel is rounded within the quarters. The small and irregular feature that is depicted (but not annotated) on the centreline at 60% of the length (from the bow) cannot be identified, but need not, perhaps, be original.

On the basis of this drawing, the form of the vessel was dissimilar-ended and the McGrail morphology code 1xx:2xx:33x or 44x:2xx:33x. It is unclear whether the vessel was flared or rectangular in cross-section.

Stuart 1866a, 119; Munro 1885, 80, 89; SAS MS/198.

3.  NX 4077 4699   NX44NW 4

This logboat was recorded by Lovaine after it had been revealed by shrinkage of the mud about 510m SW of Stonehouse farmsteading. It measured 18'6" (5.7m) in length and 2'7" (0.8m) in beam, and had probably been nearly reduced to a plank as it was said to be 'barely 2 inches' (51mm) 'deep'. A block of wood measuring 2' (0.6m) in length, 7" (178mm) in width and 5½" (140mm) in thickness had been inserted into the side 'to fit a hole, left probably by a rotten branch'. The form of the boat was unusually narrow (the slenderness coefficient being 7.16), and across the stern there was a transom-groove.

The same drawing that is noted under (2) appears (on the same basis) to depict this vessel under the heading of 'No. 2. lesser Canoe'. The vessel is depicted according to the same conventions and appears to be of similar form although, in this case, the bow is of more pointed form while the stern is cut more nearly square across and is shown as having a central forward projection of uncertain significance. A transverse 'Groove' is noted at the stern and was presumably intended to retain a transom, while there is a 'block inserted' (presumably a repair-feature of some sort) within the port side, adjacent to what appears to have been an area of splitting.

On the basis of this evidence, the form was dissimilar-ended and the McGrail morphology code 44cx:1xx:33x.

Stuart 1866a, 120; Munro 1885, 81, 89; SAS MS/198.

4-5.  NX 4092 4692   NX44NW 45

In 1863-4, a fourth logboat and part of a fifth were found in the bed of the loch about 470m SSW of Stonehouse farmsteading. The complete logboat measured 25' (7.6m) in length and was strengthened by what was apparently a false rib. This was described as a 'projecting cross band towards the centre left in the solid in hollowing out the inside'.

Beneath this boat, there was found 'a portion of another canoe'.

Stuart 1866a, 121; Munro 1885, 90.

34 **Drumduan**
NJ c. 56 01

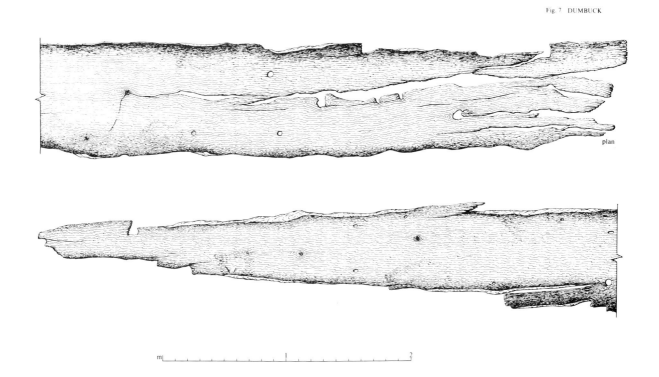

Fig. 7. Dumbuck (no. 35). Logboat. Plan view of remains. Scale 1:25.

NJ50SE 9
Grampian Region - Kincardine and Deeside District
Aberdeenshire - Aboyne and Glentanar ph.

In 1836 a logboat was found (probably during drainage operations) in 'the peat-moss at Drumduan, on the south side of Auchlossan Moss'. The area is one of afforested moorland and reclaimed peat-moss at an altitude of about 130m OD.

When found, the logboat was 'quite entire' and 'neatly formed out of a single block of oak', but it proved to be 'in an unsound state' and soon disintegrated as a result of drying and rough handling.

*NSA*, xii (Aberdeen), 1059.

35 **Dumbuck**
NS 4158 7393
NS47SW 8
Strathclyde Region - Dumbarton District
Dunbartonshire - Old Kilpatrick ph.

In 1898 a logboat was discovered about 20 yards (18m) NE of the well-known crannog that was then under excavation by Bruce and Donnelly. It was found within a 'dock-like structure' formed of 'walls...of wood and stone' which was supported by piles and linked to the crannog by a causeway. The pointed 'prow' of the boat was towards the river, and there is no indication that it was found at any great depth.

The boat immediately attracted much attention (pl. 3) largely on account of the decorated stone, bone and shell artifacts that were found within it; some of these are held in the Glasgow Art Gallery and Museum under accession number GAGM *57-96. The authenticity of these objects was doubted (most notably by Munro) and they are now universally rejected as false, but the antiquity of the logboat (and of the probable ladder that was found with it) have never been called into question. No dateable objects were found in the excavations and the date of both crannog and logboat remain uncertain, although there is no reason to doubt the conventional broad attribution to the Later Prehistoric or the Early Historic period.

On discovery, the logboat measured 35'7" (10.9m) in length over all, but before it was transferred to the museum a 'portion of the prow, which tapered to a point, and which showed two oval hand holes' was removed and lost, reducing the length to 33' (10.1m). The boat measured 4' (1.2m) in beam at the 'square' stern and was 2' (0.6m) 'deep', indicating manufacture from a half-sectioned log. There is no recorded evidence for a transom. Several 'well fitted, soft wood clamps'" (intended to hold together a split in the floor) were noted, as were several 'plugged holes' and 'marks where the seats were fitted'. The timber was identified as 'oak'.

The boat was presented by the Marchesa Chigi of Dumbuck to the Glasgow Art Gallery and Museum and is displayed under accession number GAGM *98-217a. It has been reduced by drying to a distorted, shrunken and greatly-split plank (fig. 7) which measures 9.4m in length, up to 0.9m in breadth and between 50mm and 60mm in

*Pl. 3. Dumbuck (no.35). Romantic impression by W. Milne Black of the crannog and logboat in use, which is included here as a comment on the presumed carrying capacity of an apparently-small logboat. Reproduced by courtesy of the Trustees of the National Library of Scotland from the* Scots Pictorial *for 29 October 1898.*

thickness. One end (probably the original bow) is roughly pointed while the other is greatly split but has apparently been squared across to form the stern. The timber has warped noticeably, causing the outboard parts of the floor to take on a very noticeably dished form, while the sides have split away just above the floor, leaving evidence of right-angles at the junctions. The boat was probably damaged during removal; a jagged hole near the pointed end is most probably thus explained while one of the split portions of the sides has been truncated by sawing.

Nothing can be seen of the repair-clamps that were originally noticed, or of any evidence for such internal features as seats, thwarts or fitted ribs, but a slight scrape measuring about 25mm in breadth near the edge of the floor may be the remains of a tool-mark.

The floor is pierced by six vertical thickness-gauge holes disposed in pairs along opposite sides, and also by several knot-holes; the former measure between 40mm and 75mm longitudinally and between 30mm and 60mm transversely. Distributed less regularly along the floor of the boat, but generally close to the sides, there are four uncompleted thickness-gauge holes which measure between 30mm and 45mm longitudinally by between 20mm and 30mm transversely, and penetrate the timber to a depth of between 10mm and 25mm.

A comprehensive analysis of this boat is feasible on the basis of various accounts and the incomplete remains. It was of exceptionally narrow form, having a slenderness coefficient of 8.8 and the relatively high value of 2 for the beam/draught coefficient indicates a high capacity for low-density cargoes. The displacement of 4.76 cubic metres under standard conditions falls within the middle of the normal range, while the McGrail morphology code can best be assessed as 1xx:xx1:3xx, and the form as dissimilar-ended. *August 1987.*

Murray 1898; *Glasgow Herald*, 27 January 1899; *Glasgow Herald*, 22 March 1899; Munro 1899, 438, 440; Bruce 1900, 439, 442, 448-9, 456, 460; Lang 1905, 29, 34-5, 135-6; Callander 1929, 319-20; RCAHMS 1978, 12, no. 54.

**36 Eadarloch**
NN *c.* 347 768
NN37NW 2
Highland Region - Lochaber District
Inverness-shire - Kilmonivaig ph.

In 1933 Ritchie excavated the crannog of Eilean Tigh na Slige or Eilean Ruighe na Slige in Eadarloch, which was formerly the northern extension of Loch Treig and situated in a deep glaciated valley at an altitude of about 240m OD. Excavation took place in advance of the construction of a hydro-electric power dam which has caused the loch level to rise by about 10m and covered the spit that formerly separated Eadarloch from the main body of the loch. The excavations revealed no evidence for a causeway but a probable landing-place and a mooring-post were identified on the N of the crannog.

Two 'somewhat puzzling objects of oak, which have been regarded as boats' were also found 'in the vicinity of the island'. They had been washed out from their original positions, but (in the absence of any other evidence of occupation nearby) were both probably associated in some way with the crannog. Both are on display in the West Highland Museum, Fort William.

The smaller of the two artifacts was most probably a bog butter trough and is discussed accordingly below (no. A18).

The larger object (fig. 6) was discovered after 1938 in redeposited material and was identified as probably the lower part of a logboat, although an alternative identification as a timber-skid could not be ruled out, and the two uses need not be mutually exclusive. The considerable width of the object and the presence of only a single diminutive runner may, however, argue against its being a skid, on the basis that the sides would rub excessively on uneven ground and generate considerable friction in consequence.

On discovery it measured 15'9" (4.8m) in length and 2'4" (0.7m) in breadth; the timber was identified as 'dressed oak'. A slight 'keel' was noted wrought in the solid along what was assumed to be the underside; the low 'bosses' on the other surface were identified as possible footrests.

As displayed (under accession number WHM 2251) the object is mounted vertically in a stair-well in such a way as to preclude complete examination, but there appears no reason to doubt its identification as a probable logboat. It measures (after shrinkage) 4.7m in length, up to 1.67m in breadth and between 20mm and 90mm in thickness The edges curve upwards into the sides but there is no evidence of either bow or stern structures and the boat may formerly have been longer. Twisting and splitting have occurred during drying and there is a hole around a knot at one point. Neither thickness-gauge holes or toolmarks can be identified.

The 'keel' is of roughly-square section but is slightly rounded, possibly as a result of abrasion. It runs the length of what was presumably the centre of the underside and, from its slight dimensions, would appear to have added little to the strength or wear-resistance of the boat. It may have been intended as a skewomorphic false keel modelled on that of a planked boat.

An oval rounded boss measuring about 0.17m by 0.12m and standing about 45mm high was identified about two-fifths of the way along the presumed upper surface from one end; there are the probable remains of a second at a corresponding position on the other side of the centreline. Both bosses appear to have been greatly worn down, but they were probably too small to be used as footrests and their function remains unclear.

On the basis of the available evidence, the slenderness coefficient was 6.75 which indicates a relatively narrow form. The McGrail morphology code is xxx:1xx:xxx and the form may have been that of a punt or barge. *July 1987.*

Ritchie 1942, 46-7, 57-9, 77-8.

**East Greens** See **Forfar 2**

**Eilean Ruighe na Slige** See **Eadarloch**

**Eilean Tigh na Slige** See **Eadarloch**

**37 Errol 1**
NO *c.* 26 22
NO22SE 5
Tayside Region - Perth and Kinross District
Perthshire - Errol ph.

About 1869 a logboat was found in the estuarine sandbank known as the Habbiebank in the inner estuary of the River Tay about 230 yards (250m) from the Perthshire shore and 'near to' the findspot of no. 38. No detailed record was made at the time, and there are apparently no surviving remains of the boat which was 'not in such good preservation' as that discovered subsequently. See also no. 151.

Hutcheson 1897, 265-6.

**38 Errol 2**
NO *c.* 26 22
NO22SE 4
Tayside Region - Perth and Kinross District
Perthshire - Errol ph.

In July 1895 fishermen whose net had suffered damage removed what was thought to be a tree-trunk from the Habbiebank, a sandbank situated in the inner estuary of the River Tay about 250 yards (230m) from the Perthshire shore and near to where another logboat (no. 37) had been discovered previously. The object was recognised as a

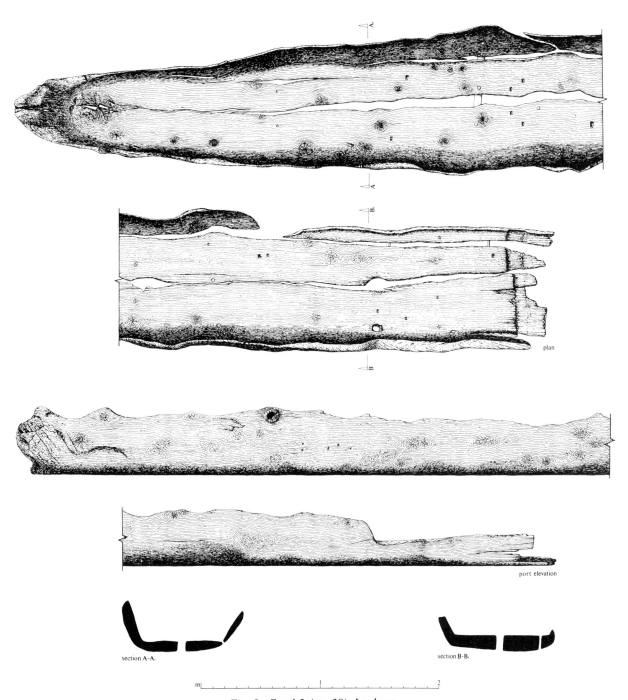

Fig. 8. Errol 2 (no. 38). Logboat.
Plan view and port elevations with sections at 36% and 85% of length from bow.
Scale 1:25.

logboat and was taken to Newburgh where it was recorded before removal to Dundee. See also no. 151.

The boat was found to measure 29'2" (8.9m) in length, 4'3" (1.3m) in beam at the stern and 3'2" (1m) in beam at a point 6' (1.8m) from the stem. The sides were about 4" (100mm) thick and up to 1'9" (0.5m) high internally above the bottom which was between 6" (150mm) and 7" (175mm) thick. The port side was almost complete but the after part of the starboard side had been lost; there were few toolmarks visible in the timber which was identified as 'oak'.

The pointed stem was seen to bear a 'rude but forcible resemblance to the head of an animal' and behind it there was noted a 'roughly semi-circular hollow' which was suggested as a possible base of a figurehead. The stern had been formed by a transom placed 10" (250mm) from the after extremity; this had measured about 3½" (90mm) in thickness but was only represented by its groove.

Within the interior of the boat, and at a point 7'4" (2.2m) forward of the transom, there was an indentation in one of the sides and across the bottom there was a 'very shallow depression'. The former was possibly intended to take a 'cross-piece of wood...tightly jammed in' and the latter a board of unknown function. A further pair of indentations in the sides was noted a further 3'7" (1.1m) forward and were taken as evidence of a seat formed of a block of wood, measuring about 6" (150mm) or 8" (200mm) square, jammed in place; no transom-groove was noted in this case. Hutcheson also suggested that these boards or blocks were possibly used to expand the boat, although the method is unusual and the low position of the point of application of force makes this unlikely.

This logboat (fig. 8) is on display in Dundee Museum and Art Gallery under accession number DMAG 69-255; the bottom and the lower part of the starboard side were inaccessible at the date of visit. It has suffered badly from splitting and is held together by iron bands. There is no sign of warping but there are numerous deep impressions which were probably made with a metal pickaxe or similar tool during recovery operations.

As reinforced for display, the boat measures 8.64m in length and up to 1.18m in beam. The floor measures up to 0.73m in breadth and the height of the sides varies around 0.3m internally and 0.5m externally. The bottom and sides measure between 90mm and 120mm and about 80mm in thickness respectively. The boat is roughly-formed of much-knotted timber and has probably been worked from a half-sectioned log. Extensive splitting has occurred along the junction of the floor with the flared sides where the timber has been left excessively thin. There are nine holes forming two transverse lines across the floor and sides about 2.3m and 5.8m respectively from the bow. These holes generally measure about 15mm in diameter and may have been drilled to receive thickness-gauges but their circular shape suggests that they have formed part of a post-shrinkage reinforcement operation. The single hole that is located on the centreline 4.4m from the bow measures 20mm in diameter and may be similarly explained. There are also numerous small holes of unknown function distributed around the boat.

The bow has probably been pointed externally in both horizontal and vertical planes but has been foreshortened by splitting and breakage; the internal form is rounded. The stern has suffered greatly from splitting and differential shrinkage, the port side being greatly reduced in height and the starboard held in place by modern ironwork. The worn and distorted remains of the groove for the (lost) transom are clearly visible; this has measured about 170mm and 120mm in breadth at top and bottom respectively, and about 60mm in depth.

On the basis of the dimensions recorded at the time of discovery, the boat was of relatively narrow form, having a slenderness index of 6.8. The beam/draught coefficient and the displacement were within the mid-range for Scottish logboats, having values of 1.3 and about 6.7 cubic metres respectively. Assuming the boat to have been worked from a whole log, as appears probable but not certain, the log conversion percentage is about 90%. The McGrail morphology code is 44a3:113:3x3 and the form is a variant of the dissimilar-ended type.

This vessel has yielded radiocarbon dates of 485 ± 40 ad (Q-3121) and 430 ± 45 ad (Q-3141), which may be calibrated to about 599 and 548 cal AD respectively. *August 1987.*

*Scottish Notes and Queries*, ix (1895-6), pt.4, 61; Hutcheson 1897, 265-72; Scott 1951, 30, n.6; Coutts 1971, 66, no. 136; McGrail 1978, i, 67; McGrail 1987a, 82; *pers. comm.* Miss C. Lavell.

### 39  Erskine 1
NS *c.* 463 721
NS47SE 45
Strathclyde Region - Clydebank or Renfrew Districts
Dunbartonshire - Old Kilpatrick ph. or Renfrewshire - Erskine ph.

In late July 1845 the greater part of a logboat was recovered from 'the north side' of the River Clyde near Erskine Ferry. It was possibly exposed by the stern-wash of river-steamersand was first noticed at low water when what was probably the forward section appeared as 'a projecting piece of wood, having the appearance of a seat'. The boat lay N-S, the stern being nearer the land and buried under about 2' (0.6m) of sand. In spite of the loss of the bow section, the boat was successfully paddled across the river to 'the ferryman's premises at Erskine' where it was recorded for a newspaper account. The remains had almost entirely disappeared by 1856.

The boat was found to measure 29' (8.9m) in length and about 5' (1.5m) in the beam at the stern and was 3'4" (1.0m) and 2'2" (0.7m) deep at the stern and amidships respectively. The bow was 'not entire' and the original length was tentatively estimated at 35' (10.7m). It had been worked from 'oak' and the widespread marks of an adze or similar tool measuring about 5" (127mm) in breadth were noted in the interior, but there was no visible charring. The sides were unusually thick, measuring 5" (130mm) in thickness near the stern while the bottom was only 2" (50mm) thick near the stern and even thinner towards the bow.

At a point 'within about three feet' (0.9m) of the foremost surviving point (within the foremost 10% of the length) there was a 'drilled hole' which measured between 2" (51mm) and 2¼" (57mm) in breadth while 'in the lower part of the vessel on each side' there was the 'usual series of holes' (numbering seven on the port side and five on the starboard) which were seen as having been intended to retain 'sticks or pins (which) must have been inserted for the feet of the rowers'. All these thirteen holes had presumably held thickness-gauges.

The boat tapered away from the stern which 'sloped from the keel backwards, just as modern boats are' on the port side while the starboard side was 'imperfect' and had been supplemented or extended with what was probably an improvised washstrake formed of a 'bent piece of wood, with a groove on its upper surface, three feet (0.9m) long, the depth of the curve between the points being eight inches (0.2m)'. This timber had been fastened to the trunk by wooden treenails, the survival of which was not noted, but for which there were four 'auger holes...one at each extremity and two towards the bottom'.

The stern itself was of transom form and comprised 'boards of oak fitted in a groove'; the survival of these is not noted but there are said to have been grooves (each measuring about 6" (0.15m) in breadth) reaching upwards to the full height of the side.

Four 'indentations' were noted set into the port quarter about 4" (100mm) below the sheerline and at intervals of 2'7" (0.8m); these probably indicated the former locations of thwarts which had been 4" (0.1m) broad. There were corresponding 'marks' on the starboard quarter and Bruce may be describing these features when he notes that this side 'showed the appearance of iron locks on each side near the stern'.

The boat displayed evidence of hard usage. The improvised washstrake on the starboard quarter is unparalleled on the port side and so was probably an inserted section to replace a loss through splitting, rather than an original construction feature. Near the port quarter there was a longitudinal split which had been repaired with a double-flanged timber which measured about 9" (230mm) by 6" (150mm) and about 3½" (89mm) in the 'middle check'; this was retained by treenails.

The presence of so many constructional and repair details on this boat makes its loss a matter for regret, particularly in view of the similarity in some respects with the Hasholme discovery (Millet and McGrail 1987) which is of similar size and complexity. On the basis of the dimensions recorded at the time of discovery, this boat was of average proportions but relatively large size having slenderness and beam/draught coefficients of about 5.8 and 1.49 respectively, and a displacement under standard conditions of about 8.2 cubic metres. The McGrail morphology code is best assessed as 44bx:2xx:xxx and the form was dissimilar-ended on account of the fitted transom.

*Glasgow Herald*, 31 July 1854; [Buchanan] 1884, 367; Bruce 1893, 18-19.


40-2 **Erskine 2-4**
NS *c.* 44 73
NS47SW 69
Strathclyde Region - Renfrew District
Renfrewshire - Erskine ph.

In 1893 there were said to be 'one or two canoes lying silted up' on the S bank of the River Clyde 'opposite Bowling' and in the area where 'what was considered to be an inverted canoe' had earlier been exposed 'for some time'.

Bruce 1893, 20.


43 **Erskine 5**
NS 449 731
NS47SW 70
Strathclyde Region - Renfrew District
Renfrewshire - Erskine ph.

In April 1977 a logboat was identified by wildfowlers after it had been revealed by river action in the inter-tidal zone about 600m NW of Erskine Hospital. It was recorded, measured and photographed, but was not recovered.
The bow (which was of rounded point form) and a parallel-sided length of 10'2" (3.1m) amidships were exposed but probing indicated that the vessel was at least 15' (4.6m) long. The internal depth was said to be at least 2'6" (0.8m). The McGrail morphology code was xxx:1xx:3xx. The form and performance coefficients cannot be ascertained.

*Paisley and Renfrewshire Gazette*, 11 March 1977; *Sunday Express*, 13 March 1977; *Paisley and Renfrewshire Gazette*, 25 March 1977; *pers. comm.* Miss H. Adamson.


44 **Erskine 6**
NS 4640 7194
NS47SE 62
Strathclyde Region - Renfrew District
Renfrewshire - Erskine ph.

In June 1977 a logboat (pls. 4 and 5) was revealed by stream erosion within the inter-tidal zone about 20m E of Bottombow Island and to the SW of the training wall. It was parallel-sided, measuring 21'4" (6.5m) over all by 2' (0.6m) in beam; the rounded point stern lay to the W and the prow (which appears from the available photographs to have been of similar form) was at the E. The 'curious wedge-shaped cutout' that was noted near the bow possibly resulted from anchor damage.

The McGrail morphology code was most probably 3xx:1xx:333, although the shape of the bow remains doubtful. The form was probably a variant on the canoe form, while the boat was apparently of exceptionally narrow shape, having a slenderness coefficient of 10.7.

The surviving drawings (redrawn as fig. 9) confirm the suggested form of the boat, but were apparently made after it had been reduced to a length of about 5.5m. The 'oarlock', or possible crutch-support, that is noted at a point about one-third of the way along one side of the boat is unparalleled in Scotland, although evidence for propulsion by rowing has been noted in England (McGrail 1978, i, 320-1). In the absence of further evidence, the significance of this feature remains unclear.

The boat was recovered in four sections (starboard quarter, stern, bow and centre section) but it disintegrated after a timber sample had been taken which yielded a radiocarbon date of 45 ± 50 bc (GU-1016), which may be calibrated to about 4 cal AD.

*Pers. comm.* Miss H Adamson.


## 45 Falkirk
NS87NE or NS88SE
Central Region - Falkirk District
Stirlingshire - Falkirk, Grangemouth or Larbert ph.

Forsyth notes that 'a complete boat' was found near Falkirk 'some years' before 1805; it was found 'five fathoms deep in the clay'. He is probably describing the discovery of a logboat in the same extensive area of estuarine carseland as that from the River Carron (no. 148); the considerable depth of the discovery (9.1m) suggests a similarly early date.

Wilson locates this discovery more specifically 'in the immediate vicinity of Falkirk' but states no evidence for doing so.

Forsyth 1805-8, iii, 419; Wilson 1863, i, 47; Munro 1898, 272; NMRS dataset NS88SE 89.


## 46 Finlaystone
NS *c.* 36 74
NS37SE 26
Strathclyde Region - Renfrew or Inverclyde District
Renfrewshire - Erskine or Kilmacolm ph.

In 1878 a 'portion of a canoe with some animal bones' was found 'on the Finlaystone Bank'. Although the discovery is mentioned only in a county history of Dunbartonshire, the location was probably on the S side of the River Clyde and in the vicinity of Finlaystone House, which is at NS 3645 7370.

Irving 1920, 151.


## 47 Flanders Moss
NS *c.* 62 99
NS69NW 3
Central Region - Stirling District
Perthshire - Port of Menteith ph.

In the latter part of the 18th century drainage operations and peat-digging revealed buried oak-trees, timber trackways and artifacts of various types in this extensive area of raised peat-bog at an altitude of about 15m OD; among the discoveries there was noted 'part of a ship'. The topography of this inland area renders this specific identification improbable, and the account may be describing the discovery of a logboat.

*Stat. Acct.*, xx (1798), 91; RCAHMS 1979, 27, no. 227.


## 48  Forfar 1
NO *c.*  456  509
NO45SE 134
Tayside Region - Angus District
Angus - Forfar ph.

In 1952 it was noted that a 'dugout' had been found in Forfar ninety years before; the boat was found 'near where Castle Motors Garage now stands'. It was apparently not recorded in detail, but lay in Forfar cemetery for a long time until it disintegrated.

Forfar Castle is said to have stood at NO 456 508, Castle Street extends from NO 4560 5094 to 4559 5061, and Castle Motors Garage was presumably one of the two garages that are noted on the 1972 edition of the OS map at NO 4559 5094 and 4564 5091 respectively.

The area is situated at an altitude of about 57m OD and has probably been occupied by a continuation of the Loch of Forfar to the E of the ridge upon which stands the core of the burgh itself.

*[Dundee] Evening Telegraph*, 12 November 1952; OS 1:2500 map (1972 ed.) NO 4450-4550.


## 49 Forfar 2
NO 4574 5087
NO45SE 22
Tayside Region - Angus District
Angus - Forfar ph.

On 11 November 1952 a logboat was found at East Green or Greens in the E part of the town of Forfar. The discovery was made in a depression in clayland at an altitude of about 65m OD, and probably during drainage or construction operations. The timber was identified as 'oak' and

*Pls. 4-5. Erskine 6 (no. 44). Two general views taken during recovery operations and showing the probable anchor damage. The boat was apparently turned round between the taking of the two photographs. Reproduced by kind permission of Miss H. Adamson from the collection of Glasgow Museums and Art Galleries.*

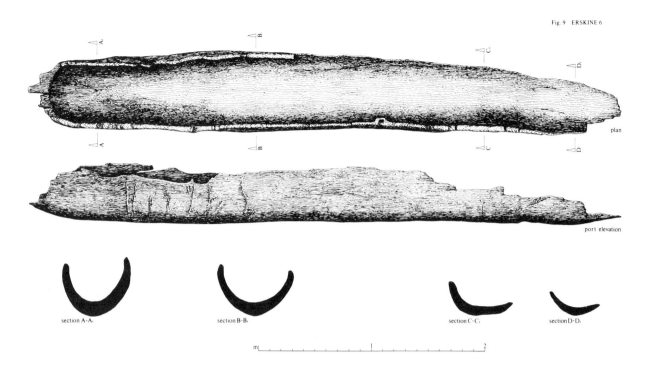

*Fig. 9. Erskine 6 (no. 44). Logboat. Plan, port elevation and four sections at roughly equidistant spacings: re-worked from excavation drawings. Scale 1:25.*

the boat was 'filled with stones'. It was taken initially to Dundee Art Gallery and Museum and, in 1986, was transferred to Angus District Museums. After conservation it was taken to Forfar Library and Museum where it is in course of display preparation under accession number ANGMAG F1987.1002.

As preserved, the remains comprise three large pieces (which have been divided by splitting along the length of the boat) and numerous smaller fragments, some of which are little more than splinters. No adequate record could be produced by drawing and the smaller fragments were not unwrapped for examination at the date of inspection. There is insufficient evidence for quantitative analysis.

The largest piece comprises the starboard section of the bottom and the lower part of the starboard side, which measures about 0.35m in height internally. The upper part of the side has been rendered highly uneven by knot-holes, splitting and saw-cuts. This piece measures 2.65m in length, and at one end it curves in towards what was apparently the bow.

The second largest piece measures 2.57m in length and includes the port side which survives to a height of 0.3m internally. At the stern there is a raised section which measures 120mm in height and 190mm in thickness.

The third piece measures 2.61m in length by 0.46m transversely and has formed the central section of the floor. The raised portion at the stern measures about 130mm high and 0.2m thick.

The size and form of the vessel cannot be established with any degree of certainty, but it has been poorly constructed from heavily-knotted timber; the minimum dimensions have evidently been at least 2.65m in length by about 0.5m transversely and 0.35m in internal depth. There are no thickness-gauge holes and the flat bottom varies between about 40mm and 110mm in thickness; the sides are about 30mm thick and have been flared outwards. It has been greatly mutilated by both warping and splitting, but part of a bow of rounded or rounded point form are still to be seen. The McGrail morphology code was probably 113:1x3:xx3, and the form possibly dissimilar-ended.

The boat has been dated by radiocarbon to 1090 ± 50 ad (Q-3143), which date may be calibrated to about 1181 cal AD. *September 1987.*

*[Dundee] Evening Telegraph*, 12 November 1952; *DES*, (1953), 5; Coutts 1971, 67, no. 138; *Dundee Courier and Advertiser*, 28 October 1994; *pers. comm.* Miss C. Lavell.

**Friar's Carse** See **Carse Loch**

**50 Friarton**
NO 1175 2192
NO12SW 24
Tayside Region - Perth and Kinross District
Perthshire - Perth ph.

In 1878 or 1879 Geikie recorded the remains of the logboat that had been found 'a number of years' before during clay-digging at the Friarton brickworks, a short distance S of the City of Perth. The exact location of the discovery was not noted but an annotation on the 1932 edition of the OS 1:2500 map places it about 90m WNW of the present harbour, and at an altitude of about 10m OD. See also no. 151.

The boat had been considerably damaged before examination when the surviving fragment was found to measure 10' (3.1m) in length and 3' (0.9m) in depth between the highest surviving and the lowest points; the dugout cavity measured 6' (1.8m) long by 2' (0.6m) deep and was 1'6" (0.5m) broad at the bottom while the highest surviving part of the side of the boat was 3" (8mm) thick. The timber was identified as 'Scotch fir' and extensive charring was noted. It was locally remembered that on discovery the boat had measured at least 15' (4.6m) in length and Geikie suggests that it had formerly measured about 3'6" (1.1m) and 3' (0.9m) in beam externally and internally respectively.

On the basis of the best available evidence for the dimensions of this boat, it was a medium-sized craft worked from a whole log and with proportional values within the mid-ranges of those normally found. The displacement was about 2.7 cubic metres while the slenderness and beam/draught coefficients were of the order of 4.3 and 1.2 respectively. The McGrail morphology code and the form of the boat cannot be ascertained on the basis of the available account.

The boat was said to have been found 'resting on its bottom' on the upper surface of the peat and sand layers that underlie the deep brickearth deposit of what Geikie termed the 'second alluvial terrace of the Tay'. This peat was laid down on the surface of former estuarine sediments during the period of low sea levels that was followed by the period of marine transgression during which the brickearth was deposited; this transgression is dated regionally to between 8400/8100 bp and 6800/6500 bp but the reported location of the discovery at an altitude of about 6.6m OD suggests that the area was inundated at a relatively late stage, probably in the later centuries of the 8th millennium bp. Assuming the account of the logboat being found at the base of the clay layer to be correct, a date within the later 6th millennium bc and in the later part of the Boreal climatic phase may be attributed to it.

Geikie 1880; Cullingford, Caseldine and Gotts 1980; OS 1:2500 map, Perthshire, 2nd ed. (1932), sheet xcviii.9; *pers. comm.* Dr. M. Armstrong and Dr. D.G. Sutherland.

51 **Garmouth**
NJ *c.* 345 630
NJ36SW 23
Grampian Region - Moray District
Morayshire - Urquhart, Speymouth or Bellie ph.

The logboat that is displayed in Elgin Museum is said on the accompanying label to have been found 'opposite Red Kirk of Speymouth', which is probably to be identified with St Peter's Church, Essil, at NJ 3395 6346. A contemporary newspaper account notes its discovery 'while clearing the river of obstructions opposite the farm of Newton' in January 1886. Newton farm is at NJ 339 624 and the boat was probably revealed by natural erosion in the braided lower course of the River Spey in the vicinity of Essil Pool (NJ 345 630). It was taken initially to Tugnet farmsteading (NJ 349 653) and donated to the museum in 1916. See also no. 74.

When discovered, the logboat was found to measure 16' (4.9m) in length by 2'6" (0.8m) and 1'10" (0.6m) in breadth at the stern and at a point 4' (1.2m) from the bow (some 25% of the length of the boat) respectively. It was 'in good preservation' and the timber was identified as 'first-class black oak'. The boat was said to be 'finely moulded, especially at the bow, where it curves upwards and shows part of the sides adhering yet'. The stern was described as 'coble fashion viz. square' and survived to a height of 4" (100mm). The bottom was 'entire' and varied between 2" (50mm) and 4" (100mm) in thickness. Two 'treenails' seen 'in the prow' and another three 'at two different places' were taken to indicate the location of seats. A probable 'plug hole' was also noted.

As displayed under accession number ELGNM: 1916.2, the boat (fig. 10) measures 4.85m in length by up to 0.76m transversely and tapers towards the bow. It has been worked from timber which is generally free of knots, and is in good condition. The greater part of the sides (which have probably been flared) are lost but there has been little splitting and the timber bears a marked polish. The stern has been warped slightly downwards and the underside (which could not be examined in detail at the date of inspection) appears to have a slight chamfer at the edges.

The bottom of the boat measures between 50mm and 70mm in thickness and set into it there are fourteen thickness-gauge holes which measure between about 20mm and 40mm in diameter. Most of them are arranged in twos or threes across the boat; two of them do not completely pierce the timber and one of them retains its wooden plug. The possibility that at least some of these numerous holes were for the attachment of fitted ribs cannot be ruled out.

Situated amidships, there are two slightly larger depressions which have rounded bottoms and appear different from the thickness-gauge holes; these have possibly been intended as the locations for pillars to support a thwart.

The bow has been formed as a rounded point, and at the rectangular stern there are the worn and ill-defined remains of a transom-groove which survives over a length of 0.5m and measures 35mm in breadth by 5mm in depth at greatest.

On the basis of the dimensions noted at the time of discovery, the slenderness coefficient of the boat is 6.4. The McGrail morphology code is 44bx:2x3:323 and the form is dissimilar-ended. *September 1987.*

*Moray and Nairn Express*, 30 January 1886; MS. note

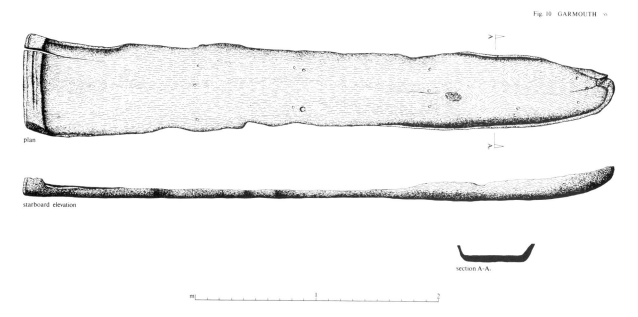

*Fig. 10. Garmouth (no. 51). Logboat. Plan, starboard elevation and section at 20% of length from bow. Scale 1:25.*

by H.B. Mackintosh in Elgin Museum Scrapbook 2.

## 52  Gartcosh House
NS 7002 6775
NS76NW  4
Strathclyde Region - Strathkelvin District
Lanarkshire - Cadder ph.

About 1892 a logboat was discovered during the digging of the foundations for Gartcosh House, which is situated in undulating clayland at an altitude of about 80m OD. No record was made of the boat, which is apparently lost.

NMRS dataset NS76NW 4.

## 53-7  Glasgow, Clydehaugh 1-5
NS *c*. 575 648
NS56SE  6
Strathclyde Region - City of Glasgow District
Lanarkshire - Govan ph.

The five logboats that are recorded from 'the lands of Clydehaugh' were found in 1852, probably during the construction of General Terminus Quay. None of them appears to have been preserved and the McGrail morphology codes cannot be ascertained on the basis of the surviving accounts.

1. In February of that year a logboat was discovered at a depth of more than 16' (4.9m) and at a location 62' (18.9m) 'back from the ancient lip of the river'. This was probably the 'very fine specimen of a single-tree canoe found at Clydehaugh' that Buchanan hoped to see displayed in the Underwriter's Hall. In the event, it was taken to Stirling's Library and subsequently to the Glasgow Botanic Gardens, where it disintegrated.

The boat was of dissimilar-ended form, measuring 12' (3.7m) in length by 2'5" (0.7m) in beam and 2'6" (0.8m) in depth. The stern was solid and the bow 'snout-like', while an internal projection at the midpoint probably held one end of a transverse seat. A curve worn in the nearby upper side (which probably resulted from rubbing by ropes) and three internal projections forward, may indicate the use of fishing nets, although there are no apparent parallels for this.

On the basis of the dimensions recorded at the time of discovery (and assuming the full height of the sides to have been recovered) this was a small, short and beamy boat with a comparatively deep draught, which had probably been worked from a whole section of log. The slenderness and beam/draught coefficients were 4.95 and 0.97 respectively, and the displacement was about 1.25 cubic metres.

2. A second logboat was found in May about 45m from (1). It lay tilted to starboard and the stem was embedded steeply into the gravel. The boat measured 14'10" (4.5m) in length, 2' (0.6m) in beam and 1'2" (0.4m) in depth. The bow was of similar form to that of (1) and a horizontal groove indicated the location of the former transom. Five probable thickness-gauge holes, each measuring about ¾" (20mm) in diameter, were noted; three of them pierced the starboard side well aft, the fourth was in the corresponding part of the port side and the fifth in the centre of the floor. On the supposition that the full height of the side has not survived, it is possible to calculate only the slenderness index, which was 7.4. The form was narrow and dissimilar-ended.

*Pl. 6. Glasgow, Clydehaugh 4 (no. 56) and Glasgow, Clydehaugh 5 (no. 57). These boats are depicted on the left and right sides respectively of a watercolour, dating from 1852, by A. McGeorge. This view probably idealises the form of the boats and exaggerates their major dimensional characteristics, but it is invaluable for their depiction of the fittings. Reproduced by kind permission of Miss H. Adamson from the collection of Glasgow museums and Art Galleries.*

Beneath the stem there was found a probable repair-patch of lead, measuring about 8" (203mm) by 5" (127mm) and pierced, probably to take seven nails; it had 'evidently not been affixed to any part of the canoe'. This object was donated to the National Museum of Antiquities of Scotland (since incorporated into the Royal Museum of Scotland), but can no longer be identified in the collections under the recorded accession number NMS MP 291.

3. In August a further three logboats, (3) - (5), were found 'close together, and within a few yards' of (2). (3) was a 'small vessel, with a snout-like bow' and in poor condition.

4. The most impressive vessel of the group measured 14' (4.3m) in length, 4'1" (1.3m) in beam and 1'11" (0.6m) in depth; it was dug from oak with the use of edged tools (including a large saw) and the stern closed with a three-piece jointed transom. Left in the solid there were two projections which had supported a midships thwart and two possible semi-circular footrests which may suggest that the craft was propelled by rowing. Near the bow there was a thickness-gauge or drainage hole which measured about 3" (76mm) in diameter and held an oak bung 'as thick as a man's wrist, and nearly a foot (0.3m) long' which was itself pierced by a circular eye, presumably to take a thong which fastened it to the inside of the boat. Several irregularly-placed thickness-gauge holes were noted aft, and within the boat there was found a piece of wood about 3' (0.9m) long, heavily abraded and pierced by circular holes stopped with wooden plugs; this was probably part of the upper side, or possibly of a wash-strake. This vessel is apparently the left-hand boat in the 1852 painting by McGeorge (pl. 6).

It is uncertain whether the full height of the sides survived, but on the assumption that they did, this was a small, short and beamy boat which had been worked from a half-section of log and was well-adapted to the carriage of bulky goods. The slenderness and beam/draught coefficients were 3.44 and 2.14 respectively, and the displacement was about 1.85 cubic metres.

5. The last discovery in the group measured 10' (3.1m) in length, 3'2" (1m) in beam and 1' (0.3m) in depth. Both bow and stern were 'sharp' and there were numerous thickness-gauge holes. A slanting indentation at the bow measured over 1' (0.3m) long and about 2" (50mm) broad and was intended to receive 'some longish four-cornered object' (possibly a splash- or washboard). Low down in one side a hole had been covered by an external wooden repair-patch measuring about 1' (0.3m) square and fastened by pegs at the corners. This vessel is probably the right-hand boat in the 1852 picture by McGeorge (pl. 6)

*Pl. 7. Glasgow, Hutchesontown Bridge (no. 59). Bow view of forward section as currently displayed. Reproduced by kind permission of Miss H. Adamson from the collection of Glasgow Museums and Art Galleries.*

who depicts a splashboard or partial washstrake retained by prominent pegs or treenails around the forward portion of the boat.

This boat was probably of canoe form, although owing to the evident loss of the upper sides it is possible to calculate only the slenderness index, which, at 3.14, indicates a relatively beamy shape. This may be in consequence of the shortness of the craft, as any boat with a beam of less than 1m would appear to be of little practical utility.

Buchanan 1854b, 213; Duncan 1883, 122-3; [Buchanan] 1884, 349-53; Paton 1890, 14-15; *PSAS*, xl (1905-6), 48.

**58 Glasgow, Drygate Street**
NS *c*. 602 652
NS66NW 16
Strathclyde Region - City of Glasgow District
Lanarkshire - Glasgow ph.
What was probably a logboat was discovered a 'number of years' before 1848, near the former Duke Street prison. The circumstances of the discovery are unknown, but the location was at an altitude of about 25m OD and about 1.2 km NE of the present river bank.

[Buchanan] 1848, 170.

**59 Glasgow, Hutchesontown Bridge**
NS *c*. 597 641
NS56SE 16
Strathclyde Region - City of Glasgow District
Lanarkshire - Glasgow ph.

This logboat was revealed in January 1880 when the level of the River Clyde was lowered by the removal of a weir; it was embedded in a clay layer on a small island 'immediately above the Hutchesontown Bridge' and 'almost immediately opposite Nelson's Monument on the Green'.

The remains of the boat comprised the forward section and one side, and measured about 24' (7.3m) in length and about 3'6" (1.1m) in beam. Three notches noted were 'evidently intended to receive the seats of the rowers' and in the bottom of the boat there was a 'raised bar' which was thought to be a footrest but was more probably a false rib, whether detached or left in the solid.

The boat suffered further damage when it was removed two months later, and is currently in course of display preparation at the People's Palace Museum, Glasgow Green, under accession number GAGM 91-69. The timber is in good condition and has a slightly polished appearance, but the existence of areas of light-toned wood may indicate former conservation treatment. There are few knots but the boat has suffered considerably from splitting and both sides are held in place by iron reinforcement.

After shrinkage, the remaining section of the boat (pl. 7 and fig. 11) measures 2.6m in length over all and up to 0.6m in beam; as now positioned the sides survive to a height of 0.32m above the flat floor. The bottom is about 40mm thick and the sides vary between about 10mm and 40mm in thickness, being thicker near the stern. There are no thickness-gauge holes, but near the starboard bow there are marks made by an adze or chisel-like implement which measured about 32mm in breadth and was struck in a forward direction. On the evidence of the surviving remains, the slenderness coefficient was apparently about 6.9.

The bow is rounded and the sides have probably been slightly flared. The stern has been closed by a transom which is now missing; its groove measures about 15mm in breadth and about 5mm in depth. The McGrail morphology code is 44ax:1xx:533 and the form was a variant of the dissimilar-ended category. *August 1987.*

Duncan 1883, 123-8.

## II Gazetteer of Logboat Discoveries

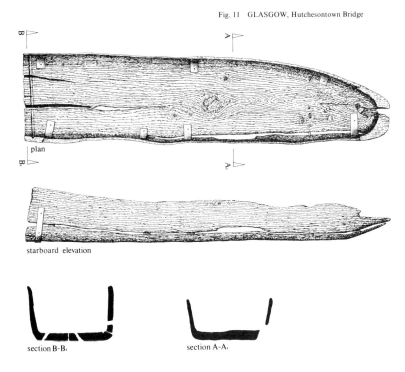

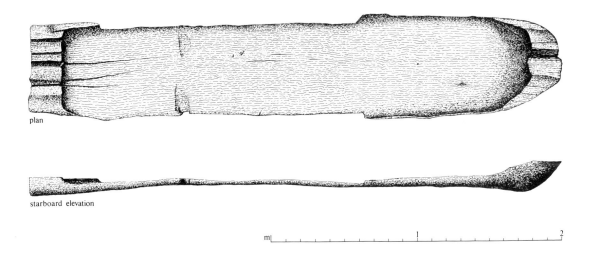

*Fig. 11. Glasgow, Hutchesontown Bridge (no. 59). Logboat. Plan, starboard elevation and sections at 45% of length from bow, and at the stern. Scale 1:25.*
*Fig. 12. Glasow, Rutherglen Bridge (no. 63). Logboat. Plan and starboard elevation. Scale 1:25.*

### 60  Glasgow, London Road
NS *c.* 596 648
NS56SE  37
Strathclyde Region - City of Glasgow District
Lanarkshire - Glasgow ph.

In July 1852 a logboat was revealed during sewerage works in London Road (then named London Street) and 'near the site of the Trades' Land'. The area is situated at an altitude of about 30' (9m) OD in the upper river terraces of the Clyde.

Buchanan notes that the boat was embedded vertically, the prow being uppermost. Wilson states that it measured 18' (5.5m) in length and was 'built of several pieces of oak, though without ribs' but he is probably confusing this discovery with that of the probable composite vessel from Bankton (no. A23).

[Buchanan] 1848, 170, 171; Wilson 1851, 35-6.


### 61  Glasgow, Old St Enoch's Church
NS 589 649
NS56SE  4
Strathclyde Region - City of Glasgow District
Lanarkshire - Glasgow ph.

In 1780 foundation work in the area now occupied by St Enoch Square revealed a logboat at a depth of about 7.6m and about 150m NNE of the present river bank. The boat lay horizontally and within the forward part there was found a jadeite axe (which is now in Glasgow Art Gallery and Museum under accession number GAGM A8931).

[Buchanan] 1848, 170; Wilson 1851, 34-5; [Buchanan] 1884, 343; Smith 1963, 144-5, 167, no. 54; Murray 1994, 97, 102, 103, no. 54; *pers. comm.* Dr. C. Batey.


### 62  Glasgow, Point House
NS *c.* 555 660
NS56NE  16
Strathclyde Region - City of Glasgow District
Lanarkshire - Glasgow ph.

In December 1851 earthmoving operations to the NW of the mouth of the River Kelvin revealed the remains of a logboat, which had probably been worked from a whole section of log. The boat measured 12' (3.7m) in length, 2' (0.6m) in beam and 1'10" (0.5m) in depth; part of the groove that held the transom was still to be seen.

Among the manuscripts of the Society of Antiquaries of Scotland there is an unsigned, undated and possibly idealised sketch (in pencil and pastel crayon) entitled 'Ancient Canoe found near Point House Ferry, on the Clyde, near Glasgow, December 1851, in possession of the River Trustees'. It depicts what appears to be a vessel of dissimilar-ended form, surviving to nearly the full height of the side on the starboard bow (which is of rounded form) but has been nearly reduced to a flat plank elsewhere. Further aft, the sides appear to rise at a sharp angle from the flat bottom; the form of the stern itself cannot be discerned (the location of the transom-groove being obscured), but the vessel does not appear to narrow at the quarters. The artist has provided no scale as such, but comparison with the standing figure alongside appears to confirm the dimensions cited above.

The fate of this vessel is not recorded but it can no longer be located. The McGrail morphology code appears to have been 44A1:1x1:222.

Buchanan 1854a, 44; Duncan 1883, 122; [Buchanan] 1884, 348; SAS MS/198.


### 63  Glasgow, Rutherglen Bridge
NS *c.* 606 630
NS66SW  9
Strathclyde Region - City of Glasgow District
Lanarkshire - Rutherglen ph.

The logboat that was discovered in the River Clyde 'close to Rutherglen Bridge' and 'a few years' before 1880 is now in store at Glasgow Art Gallery and Museum under accession number GAGM 80-8. The lower surface was inaccessible at the date of visit.

This boat has been reduced to a worn plank from which project the lower parts of the two ends. The timber is almost entirely knot-free but is now soft in texture and the edges have been rounded through wear. Splitting has occurred around the bow (part of which has broken off) and the centre section is warped upwards.

The remains (fig. 12) measure 3.66m in length by up to 0.7m transversely and the junction of the bottom with the (missing) sides is rounded with no chamfer. The bottom measures about 40mm in thickness and has been fashioned without the use of thickness-gauges. Within the boat there are numerous toolmarks. The stern is solid and square, and the bow is rounded in all three planes.

The boat has been exceptionally beamy with a slenderness coefficient of 2.56. The McGrail morphology code is 11x:1xx:222 and the form is dissimilar-ended. *October 1987.*

Duncan 1883, 123.


### 64-8  Glasgow, Springfield 1-5
NS *c.* 578 647
NS56SE  5
Strathclyde Region - City of Glasgow District
Lanarkshire - Govan ph.

In 1847-9 five logboats were discovered when the channel of the River Clyde was being widened at Springfield and the quay of that name constructed.

1. In the autumn of 1847, this logboat was discovered in a layer of sand at a depth of 17' (5.2m) and 'nearly opposite the western termination of the Broomielaw Quay'. It was recorded as being of 'oak' and measured over 11' (3.4m) in length by 2'3" (0.7m) in breadth and 1'3" (0.4m) in depth. The pointed bow was better preserved than the stern and the groove for the transom was recognised. This is apparently the left-hand boat in the 1847 painting by McGeorge (pl. 8).

The boat is housed in the reserve collection of the Royal Museum of Scotland under accession number NMS IN 1, together with several unprovenanced fragments of timber which may have formed part of it. It is in moderately good condition and has few knots, although iron reinforcing bands have been attached to the sides on account of the pronounced longitudinal splitting.

After shrinkage, the boat (fig. 13) measures 3.15m in length over all, 0.54m in beam at the stern, 0.46m in beam amidships, and about 0.3m in depth. The floor is about 70mm thick and the sides rather thinner. The groove for the transom is to be seen in the extensively-split floor but the quarters are both missing; the groove itself measures about 70mm wide and 30mm deep but the edges are not well-defined. The centre-section tapers towards the bow and the exterior of the port side is remarkably smooth, possibly as a result of river abrasion. The after section of the starboard side is missing, as is most of the port side. The sheerline had probably been lost before the discovery of the boat.

On the basis of the dimensions originally recorded, the boat may have been worked from a half-sectioned log. The slenderness coefficient was 4.86, the McGrail morphology code was 44a1:2x1:122 and the form dissimilar-ended.

This boat has been radiocarbon-dated to 1065 ± 50 bc (GrN-19280), which determination may be calibrated to about 1161 cal AD. *May 1987.*

2. In October 1848 the discovery of a second logboat was recorded 'about 400 yards further up the river...and exactly opposite the well-known dock of Mr Napier'. It was in the same sandy layer as (1) at a depth of 19' (5.8m). This is probably the boat that Edwards designates as Clyde 7.

Although damaged during discovery, it was found to measure 19'4" (5.9m) in length, 3'6" (1.1m) in beam at the stern, 2'9½" (0.9m) in beam amidships, and 2'6" (0.8m) in depth. The timber was identified as 'oak'. The prow was 'rather neatly formed, with a small cutwater' and a hole nearby was possibly intended for a painter or boatrope. The stern was closed by a transom and, left in the solid, there were rests for thwarts amidships and at the stern. A row of holes along the sheerline may be evidence for an outrigger or washstrake. On the evidence of these holes this is the right-hand boat in the 1847 painting by McGeorge (pl. 8) although that illustration does not depict the cutwater at the bow (which was possibly obscured by the angle of view) and the date of the painting apparently precedes (and casts doubt upon) the recorded date of discovery.

The remains of this boat (fig. 14) are displayed at the Hunterian Museum, University of Glasgow under accession number HM A.32. The transom, the starboard side and the forward part of the port side have been lost and the remainder of the port side has been reduced to a height of about 0.25m above the floor. The boat measures 5.56m in length by up to 0.81m in beam; the sides and bottom measure about 50mm and 80mm in thickness respectively. There is little evidence of either warping or splitting and the timber is generally free of knots, but the boat has been worn and distorted around the port quarter.

The bow survives to a height of 0.4m above the lowest part of the boat and has been formed on plan as a rounded point with a projecting prow which measures 80mm in length along the centreline and 65mm in width. In elevation, this feature tapers into the lower part of a rounded bow.

The greater part of the boat is parallel-sided on plan and flared in transverse section, the angle of flare being more pronounced on the exterior. The boat becomes broader from amidships (where it measures about 0.75m in beam) towards the stern, where the equivalent measurement is about 0.8m. Three thickness-gauge holes, and a possible uncompleted fourth, penetrate the timber but form no recognisable pattern; each measures about 25mm in diameter and the two nearest the centre of the boat retain the gauges themselves.

The stern has been squared and closed by a transom, the groove for which measures about 50mm in breadth and 25mm in depth, and is clearly visible for a length of 590mm in the starboard half of the boat; what are probably the marks of a chisel or similar metal tool are to be seen. Forward of the groove there is an ill-defined raised section which measures about 90mm in breadth and 20mm in height.

On the assumption that the full height of the side still survived, at least in part, when the boat was discovered, it had probably been worked from a whole section of log. The slenderness and beam/draught coefficients were 5.5 and 1.4 respectively, and the displacement under standard conditions was about 2.9 cubic metres. The McGrail morphology code is 44a1:1x1:322 and the form dissimilar-ended. *October 1987.*

3. The logboat discovered 'very soon' after (2) and 'within a few yards of it' was 'larger, and of a more rude description'. It suffered extensive damage during discovery but the largest fragment of 'oak' measured 9'2" (2.8m) in length, 3'6" (1.1m) in breadth and 1'6" (0.5m) in depth. The bow was 'like the snout of the modern coble'. In the bottom there were several circular indentations which were probably intended to receive the heels of pillars supporting the thwarts, and what was probably a thickness-gauge hole was found plugged with 'cork'. The incom-

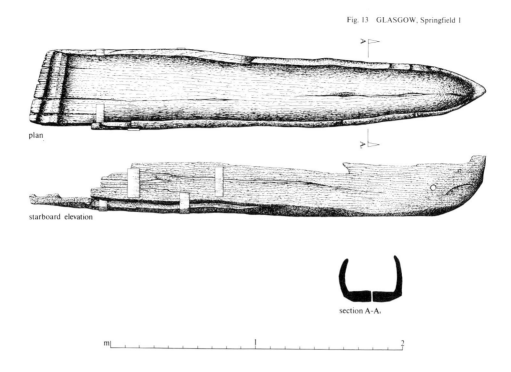

*Fig. 13. Glasgow, Springfield 1 (no. 64). Logboat. Plan, starboard elevation and section amidships. Scale 1:25.*

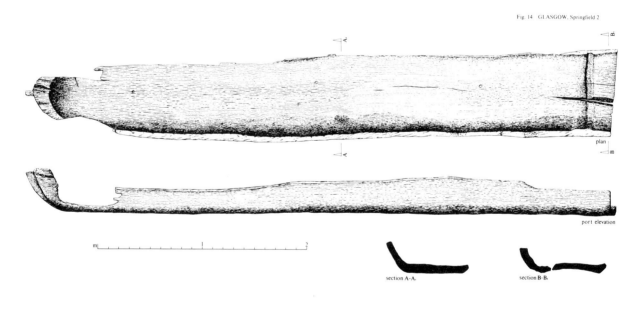

*Fig. 14. Glasgow, Springfield 2 (no. 65). Logboat. Plan, starboard elevation and sections amidships and at stern. Scale 1:25.*

II *Gazetteer of Logboat Discoveries* 43

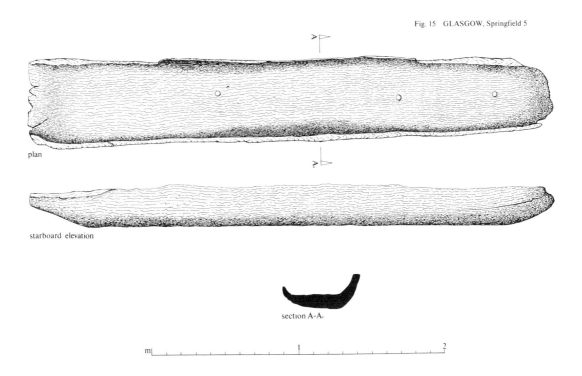

*Fig. 15. Glasgow, Springfield 5 (no. 68). Logboat. Plan view, starboard elevation and sections amidships. Scale 1:25.*

*Pl. 8. Glasgow, Springfield 1 (no. 64) and 2 (no. 65).*
*These boats are depicted on the left and right sides respectively of a watercolour, dating from 1847, by A. McGeorge.*
*Reproduced by kind permission of Miss H. Adamson from the collection of Glasgow Museums and Art Galleries.*

plete nature of the discovery precludes quantitative analysis of this boat, which was subsequently destroyed, and the form cannot be assessed on the evidence of the surviving account.

4. The logboat that was found on 7 September 1849 at a depth of 'rather more than' 20' (6.1m) was the 'most rude' of the group and was apparently extensively charred; the bottom was 'left very thick'. The boat was 'bluff' at each end and measured 13' (4m) in length, 2' (0.6m) in beam and 1' (0.3m) in depth. On the assumption that the sides were incomplete, only a slenderness coefficient of about 6.5 can now be established.

Two fragments of perforated wood found with the boat were interpreted as the remains of outriggers but were more probably pieces of the sheerline or, possibly, washstrake. The boat (which was possibly of punt or barge form) was subsequently destroyed.

5. The last logboat in the group was found soon after (4) and was taken to the Andersonian Collection. On discovery, it measured 11'10" (3.6m) in length; the beam measurements were 2' (0.6m) at the stern, 1'10" (0.6m) amidships and 1'8" (0.5m) forward. Disintegration along the sheerline had reduced the depth to only 7" (178mm) and both bow and stern were of rounded form.

This boat (fig. 15) is in store at Glasgow Art Gallery and Museum under accession number GAGM 02–73xf. It is attached by iron bolts to a pallet so that the bottom is inaccessible. It has been worked from high quality timber with no knots and has suffered slightly from splitting but not at all from warping, and has been greatly worn down so that the greater part of one side has been lost.

As preserved, the boat measures 3.61m in length by up to 0.55m transversely and the starboard side survives to a maximum height of 0.12m. The bottom measures between about 80mm and 100mm in thickness, and is generally thicker on the starboard side. No thickness-gauge holes have been cut in the boat but the marks of an adze or similar metal tool are clearly visible both inside and (more extensively) outside the boat. The small and deep indentation that is situated near the centreline about 1.35m from the stern was possibly made with a spiked implement during recovery operations.

The end of the boat that is square-cut and externally raised is assumed to be the stern; the other extremity is rounded. On this basis, and on the understanding that the sides nowhere survive to their full height, the slenderness coefficient is about 5.9, the McGrail morphology code is 121:1x1:221, and the form is dissimilar-ended. *October 1987*.

[Buchanan] 1848, 168–9; Stuart 1848, 49; *Archaeologia Scotica*, v (1873–90), pt. 3, 61; [Buchanan] 1884, 345–8; Riddell 1979, 6, pl. 4.

**69 Glasgow, Stobcross**
NS *c.* 56 65
NS56NE 97
Strathclyde Region - City of Glasgow District
Lanarkshire - Glasgow ph.

In 1875 a logboat was discovered during the construction of Stobcross Quay and in 'a deposit of sand and gravel' at a level about 8' (2.5m) below the high water mark and 3' (0.9m) above low water. It measured about 20' (6.1m) in length and a piece of antler was found in it. The boat subsequently disintegrated and the remains were destroyed by burning.

Paton 1890, 15-16.


**70 Glasgow, Stockwell**
NS *c.* 592 648
NS56SE 15
Strathclyde Region - City of Glasgow District
Lanarkshire - Glasgow ph.

In 1824 a logboat was revealed during sewerage works (and probably within the area subsequently occupied by St. Enoch station) about 200m NNE of the present river bank and at an altitude of about 7m OD. It was apparently not recorded in detail.

[Buchanan] 1848, 170.


**71 Glasgow, Tontine**
NS 5962 6491
NS56SE 19
Strathclyde Region - City of Glasgow District
Lanarkshire - Glasgow ph.

In 1781 foundation work revealed a logboat on the N side of Glasgow Cross, about 410m NNE of the present river bank and in a substratum of laminated clay. The depth of the discovery is nowhere recorded, but the area is situated at an altitude of 32' (9.8m) OD in the upper river terraces.

The Tontine Hotel was built in about 1737-60 and stood next to the former tolbooth which was redesigned in 1814. What became the Tontine Building was acquired by the Tontine Society in 1781 and the logboat was presumably found during subsequent reconstruction works.

[Buchanan] 1848, 170; Wilson 1851, 35; *pers. comm.* Mr. G. Stell.


**72-3 Glasgow, Yoker 1-2**
NS *c.* 51 68
NS56NW 9
Strathclyde Region - City of Glasgow District

Dunbartonshire - Old Kilpatrick ph.

In 1863 dredging operations revealed two logboats on the NE bank of the River Clyde 'nearly opposite Renfrew' and at a depth of 14' (4.3m); they were subsequently broken up. Both were of 'oak' and had 'closed' sterns; neither had a cutwater.

1. The larger boat measured about 25' (7.6m) in length.

Buchanan 1883, 76, 77.


**Glen of Craigston** See **Craigsglen**


**74 Gordon Castle**
NJ 35 59
NJ35NE 11
Grampian Region - Moray District
Morayshire - Bellie ph.

About 1886 a logboat was found in the Bog of Gight, an area of reclaimed marshland to the S of Gordon Castle at an altitude of about 20m OD. Although this discovery has been equated with the logboat that is in Elgin Museum, museum records indicate that the boat there displayed is that from Garmouth (no. 51). The Gordon Castle example was 'destroyed sometime at the Fochabers Stocks'.

*The Northern Scot*, 21 October 1916; MS. note by HB Mackintosh in Elgin Museum Scrapbook 2.


**75-6 Kilbirnie Loch 1-2**
NS 323 535
NS35SW 5
Strathclyde Region - Cunninghame District
Ayrshire - Kilbirnie ph.

In 1868 a crannog, a causeway and up to four logboats 'in a less or more entire condition' were revealed at the SW corner of Kilbirnie Loch, which is situated in the industrialised claylands of North Ayrshire at an altitude of about 30m OD. There is no evidence for the direct association of the logboats with the crannog. The dumping of furnace-slag from the adjacent ironworks had caused the lake-bed sediments to be displaced upwards and towards the NE so that the archaeological remains were exposed above the surface of the water. Although some of the published accounts are incorrectly orientated, what are probably the crannog and its causeway are depicted on the 1st edition of the OS 6-inch map at NS 3238 5356.

Two of the boats were recorded in some detail, although the incomplete nature of the remains precludes quantitative analysis or the establishment of McGrail morphology codes.

1. The most complete of these boats was found about 20' (6.1m) N of the crannog. It measured about 18' (5.5m) in length, 3' (0.9m) in breadth and 'close on' 2' (0.6m) in 'depth'; a further length of about 2' (0.6m) had been lost from the bow. The stern was 'square'.

A hole in the bottom of the boat had been repaired, and the 'thin plate or piece of metal' that was found in it may have been a repair-patch. The function of the 'wooden pins' that were also found is unclear.

Also found in the boat (but not necessarily associated with it) there were a bronze ewer and a tripod pot which are held in the collections of the Royal Museum of Scotland. The boat itself disintegrated rapidly on exposure to air.

2. Part of a second logboat was subsequently found 'close by the island' (and on its NE side) when some of the lake-bed sediment was being dug out for agricultural fertiliser; its fate is not recorded. This boat was worked from 'oak' and measured 5 or 6' (1.5-1.8m) in length by about 2'2" (0.7m) in maximum beam; the floor was between 2½" (65mm) and 3" (76mm) thick and the sides survived to a maximum height of 6" (150mm).

OS 6-inch map, Ayrshire, 1st ed., sheet viii; Cochrane Patrick 1872, 385-6; Love 1876, 287.


**77 Kilbirnie Loch 3**
NS 3278 5373
NS35SW 8
Strathclyde Region - Cunninghame District
Ayrshire - Kilbirnie ph.

In April 1930 a further logboat was found about 410m NE of the findspots of nos. 75 and 76, and as a result of the same processes. It was buried in fine mud, probably after sinking in deep water.

When discovered, it measured about 11'6" (3.5m) in length over all and up to 1'11¾" (0.6m) in breadth, and was 1'6" (0.5m) 'deep'. The marks of an 'axe' were noted and amidships there was a recess which was probably intended to receive a thwart. Two holes piercing the starboard quarter were identified as probably intended to retain a pair of rowlocks or thole-pins; there were no corresponding perforations in the port side.

Immediately after its discovery the surviving remains, which comprise the bottom and the lower parts of the sides and ends, were taken to the Glasgow Art Gallery and Museum where they are in store under accession number GAGM 30-7, having been bolted to a pallet which renders the bottom inaccessible. After shrinkage, the remains (fig. 16) measure 3.1m in length by up to 0.71m transversely; the starboard side survives to a maximum height of 0.28m. Extensive flaking (which superficially resembles tool-marking) has occurred near the broader end, and the timber is greatly split and knotted.

The boat has been crudely constructed without the use

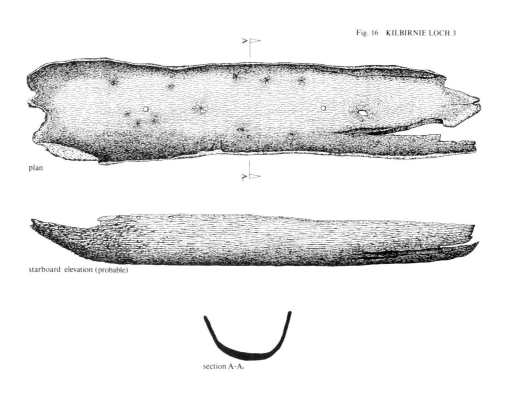

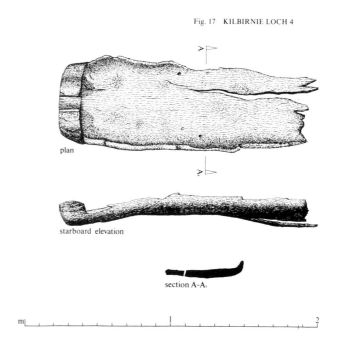

*Fig. 16. Kilbirnie Loch 3 (no. 77). Logboat. Plan, elevation of (probable) starboard side and section amidships. Scale 1:25.*
*Fig. 17. Kilbirnie Loch 4 (no. 78). Logboat. Plan, starboard elevation and section at roughly midpoint of surviving remains. Scale 1:25.*

of thickness-gauges and is noticeably assymetrical; the sides and the bottom measure about 30mm and between 30mm and 50mm in thickness respectively. Within what was probably the starboard side at a point 1.2m from the broader end (and some 66% of the length from the assumed bow) there is a possible support for a thwart in the form of a hollow measuring 80mm square and 20mm deep which is defined on three sides by a raised portion about 70mm thick left in the solid. Towards the broader end (at 69% and 76% respectively of the surviving length) there are two oval holes which measure 40mm by 20mm and are set 0.2m apart just below the top of the surviving side; their function is unclear and there is no comparable feature in the other side.

The boat has probably been worked from a whole section of log and the two extremities are badly damaged but appear to be of rounded form; the broader was more probably the stern. The McGrail morphology code is 222:112:222 and the shape is a variant of the canoe form. *October 1987.*

Mann 1933, 139-42; McGrail 1987a, 80.

## 78 Kilbirnie Loch 4
NS *c.* 32 54
NS35SW 11
Strathclyde Region - Cunninghame District
Ayrshire - Kilbirnie ph.

In May 1952 part of a logboat was found on the W side of Kilbirnie Loch and on the property of the Glengarnock Steelworks, through whose slag-dumping operations it had been revealed. Analysis of pollen from the mud found in the interstices of the timber suggested that the logboat might be tentatively assigned to the Sub-Boreal period (pollen zone VIIb), between about 3000 and 700bc.

The surviving portion of the boat (fig. 17) was donated to Paisley Museum and is in store under accession number PM 14a-1952. It measures 1.8m in length by up to 0.62m transversely and comprises a substantial fragment of the flat bottom from which project the lower part of one end, and part of one side. The blunt form of the surviving end suggests that it was probably the stern, and the side now measures about 100mm in height. Both warping and splitting have been severe and wax-treatment has been applied.

The bottom varies in thickness between 15mm and 50mm, and among the numerous knot-holes there are four vertical thickness-gauge holes which form two pairs on opposite sides. These have evidently been bored from inside the boat as two of them do not penetrate the timber; three of them measure 25mm by 20mm and the other 25mm by 16mm. The suggestion made in the museum records that these are lacing-holes is incorrect but the possibility that they were intended to retain fitted ribs cannot be discounted. There are slight traces of what may be toolmarks.

The McGrail morphology code of this boat is 12x:xxx:xxx or (less probably) xxx:xxx:12x and the form cannot be assessed. *August 1987.*

*Paisley Museum Annual Report for 1952-3*, 14; Letter to Mrs E Grant from CH Rock, Paisley Museum and Art Galleries, dated 17/2/60, enclosing pollen analysis report by SE Durno.

## 79-80 Kilblain 1-2, and paddle
NY *c.* 024 695
NY06NW 28
Dumfries and Galloway Region - Nithsdale District
Dumfrieshire - Caerlaverock ph.

Pennant notes the discovery at different dates of two 'antient canoes of the primaeval inhabitants of the country' in a 'morass' near Kilblain. Munro confuses these discoveries with that from Lochar Moss (no. 123). The area is cultivated hilly clayland at an altitude between 10m and 35m OD. The former church of Kilblain has been located at NY 0244 6958, near the present farm of Kirkblain.

1. The vessel that was discovered in 1736 measured 7' (2.1m) in length and was found with a paddle.

2. The second vessel was examined in 1772 and found to measure 8'8" (2.7m) in length over all and 6'7" (2m) internally; its 'breadth' was 2' (0.6m) and its 'depth' 11" (280mm). It had evidently been hollowed with fire, and at one end there were 'the remains of three pegs for the paddle'. On the assumption that the full length was preserved but that the sides were incomplete, only a slenderness coefficient of about 4.3 can now be established.

Pennant 1774-6, i, 94; Wilson 1851, 30-1; Munro 1898, 273.

## 81 Kinross
NO *c.* 112 022
NO10SW 6
Tayside Region - Perth and Kinross District
Kinross-shire - Kinross ph.

About 1862 a logboat was discovered during the construction of the Devon Valley railway across an area of NE-sloping clayland at an altitude of about 125m OD near the former farmsteads of Bowton. The boat was taken to Kinross House but is now lost.

OS 6-inch map, Fife and Kinross, 1st edition (1857-66), sheet xxii; Burns-Begg 1901, 5.

**Kirkblane**  See **Kilblain**

## 82 Kirkmahoe
NX c. 972 815
NX98SE 10
Dumfries and Galloway Region - Nithsdale District
Dumfriesshire - Kirkmahoe ph.

In 1919 the remains of a logboat were found when the burn known as 'the Lake' or Lake Burn was being deepened at a point about 300 yards (275m) W of Kirkmahoe parish church. The locality is one of hilly clayland at an altitude of about 15m OD, above the haughland of the River Nith.

One end of the boat was found split into two parts, the longer of which measured 5' (1.5m) in length. The timber was identified as 'oak' and was 'greatly decayed', but the beam appeared to be about 1'6" (0.5m) and the surviving 'depth' about 5" (125mm). The form of the surviving end was a rounded point externally and a rough square internally.

Although it was intended to preserve the boat at the village hall, there is no record of its survival.

M'Dowall 1920.


**Kirriemuir II**  See **Auchlishie**


## 83 Knaven
NJ c. 89 42
NJ84SE 4
Grampian Region - Banff and Buchan District
Aberdeenshire - New Deer ph.

In 1850 a logboat was discovered during drainage operations in the Moss of Knaven, and at a depth of 5' (1.5m). It was subsequently placed under water for preservation at Nethermuir (NJ 91 43). The discovery is not noted either on the 1st edition of the Ordnance Survey 6-inch map or in the *Name Book* but the OS suggest a findspot at NJ 8933 4280, presumably on the basis of its reported discovery 'at the head of a small ravine'. The attribution of the discovery by Wilson to a date about 1830 appears to be erroneous. Enquiries in 1987 and 1992 revealed no local memory of this discovery.

The boat was 'formed out of a single oak-tree' and was 'of very rude manufacture'. It measured 11' (3.4m) in length and 'nearly' 4' (1.2m) in beam; at the suggested stern there was a 'projecting part, with an eye in it for the purpose of mooring'.

*The Gentleman's Magazine*, new series, xxxiii (January-June 1850), 197, reprinted in Gomme (ed.) 1886, i, 55-6; Wilson 1851, 38-9; NMRS MS/736/4; *pers. comm*. Niall Gregory.

## 84 Larg
NX c. 432 659
NX46NW 32
Dumfries and Galloway Region - Wigtown District
Kirkcudbrightshire - Minnigaff ph.

Mitchell notes that a logboat was found before 1863 'on the farm of Bents or Larg, not far from Kirrochtree'. The circumstances of the discovery are not stated and it does not appear to survive.

The boat measured about 11' (3.4m) long by 3'2" (1m) 'wide' and 1'8" (0.5m) 'deep' when it was inspected in the grounds of what is now Kirroughtree House Hotel (NX 4222 6604). The timber was identified as 'oak'.

The area is one of cultivated grassland and forested hills at an altitude of about 35m OD.

Mitchell 1864, 26.


## 85 Lea Shun
HY c. 660 214
HY62SE 19
Orkney Islands Area
Orkney - Stronsay ph.

In April 1887 what was possibly a logboat (pl. 9) was discovered in wet sand about 130 yards (119m) W of the Loch of Leashun, a small coastal loch which is situated near the S tip of Stronsay and is separated from the sea by an ayre. Local tradition recalls that the object was in reality a portion of a shipwreck which was accepted in error after being submitted as a hoax to Tankerness House Museum, Kirkwall where it was subsequently 'treated with salt' and displayed under accession number TH58 until it was accidentally destroyed in the 1960's.

When discovered, it measured 13'9" (4.2m) in length over all, and about 7" (175mm) in 'depth'. It had measured about 3'3" (1m) in beam, but the loss of one side (taken to be the starboard) had reduced this figure to 2'5" (0.7m). The timber was 'exceptionally well cut out' and about 1½" (40mm) thick, except at the ends where it was heavily knotted; in these areas it was roughly twice as thick. The sides were straight and the ends formed as rounded points. A photograph in the collections of the Royal Museum of Scotland shows the surviving side to be rounded into the bottom of the boat and its top to be level. The McGrail morphology code of the boat was apparently 322:112:322 and the form that of a canoe variant.

Twelve holes, each measuring between 1½" (40mm) and 1" (25mm) in diameter, pierced the timber of the boat. Seven of them were spaced along the bottom and were said to be 'evidently for holding on the keel', but were more probably thickness-gauge holes as some at least held 'plugs driven from the inside'. There were a further five plugged holes of identical function in the surviving por-

*Pl. 9. Lea Shun (no. 85). Undated photograph by John B. Russell of the remains of this boat, presumably taken at Tankerness House. Reproduced from the manuscript collection of the Society of Antiquaries of Scotland held in the Museum of Scotland (SAS Ms. 307/12). I am grateful to the Society of Antiquaries of Scotland for permission to reproduce this plate.*

tions of the sides.

The metal fittings of this boat are of particular interest. What was taken to be the bow had been strengthened with a metal 'binder' which was recessed into the inner edge of the timber over a length of 4½" (115mm). Two repair-patches of 'oak' had been inserted into the surviving side and fastened with iron nails to cover large knotted patches. One of them, measuring 9¾" (250mm) by 7½" (190mm) had been placed high on the side 5' (1.5m) from the presumed bow. The other measured 10" (250mm) by 7" (175mm) and was lower down and nearer the same end.

Cursiter 1887; Marwick 1927, 65; photograph in RMS, NMAS/MISC 370/12; *pers. comm.* Dr. R.G. Lamb and the late Mr. E. MacGillivray.

**86 Lendrick Muir**
NT *c.* 022 999
NT09NW 8
Central Region - Clackmannan District
Clackmannanshire - Muckhart ph.

The discovery of what were possibly the remains of a logboat has been recorded to the N of the River Devon and near to Naemoor House (now Lendrick Muir School) which is located at NO 0231 0033. Two 'large pieces of wood...seemingly charred, and hollowed on one side' were identified as 'probably portions of a canoe'.

The date and circumstances of the discovery are nowhere specified, but it is reported with that of a stone axe unearthed about 20 yards (18m) away, and was probably similarly found during drainage operations in 1881.

The area of the discovery is in agricultural land between about 0.5 and 1km NE of the scenic gorge at Rumbling Bridge, and at an altitude of about 135m OD. The River Devon is a small upper-course river which is unlikely to have been navigable at this point, and it is probable that the logboat (if such it was) was used on a former loch in the now-drained land around.

Murdoch 1882.

**87-8 Lindores 1-2**
NO *c.* 24 19
NO12SW 62 and NO21NW 6
Fife or Tayside Region - North-East Fife or Perth and Kinross District
Fife or Perthshire - Newburgh or Errol ph.

About 1816 two logboats were found 'in the bed of the Tay, opposite Lindores Abbey' and in the area of Cruive or Oldcruvie Bank with is noted on a map of 1866. Both boats were subsequently cut up to serve as building-lintels.

These are probably the two logboats found 'in the Tay itself' that are mentioned in an account of 1881; the 'canoe excavated at Newburgh-on-Tay' that Shearer noted in 1907 is probably one of them. So also is the 'very well preserved' logboat that a newspaper article of 1848 notes as having been found about 30 years before. See also no. 151.

1. The larger boat measured 28' (8.5m) in length and was 'quite entire'.

*Perthshire Courier*, 22 June 1848; OS 6-inch map, Perthshire, 1st ed. (1866), sheet xcix; Laing 1876, 1-2; *Proceedings of the Perthshire Society of Natural Science*, (1881-6), 30; *Transactions of the Stirling Natural History and Archaeological Society*, (1906-7), 97.

## 89 Linlithgow, Sheriff Court-house
NT 0016 7710
NT07NW 31
Lothian Region - West Lothian District
West Lothian - Linlithgow ph.

About 1860 a logboat was discovered during the digging of foundations for Linlithgow Sheriff Court-house which is built on clayland about 140m SE of the shore of Linlithgow Loch at an altitude of about 55m OD and within the area occupied by the medieval Royal Burgh. The boat was taken to the 'Antiquarian Museum, Queen St, Edinburgh' but it cannot now be identified in the collections of the Royal Museum of Scotland and there is apparently no description of it.

Macdonald 1941, xv, fn. 2.

## 90 Littlehill
NS *c*. 624 713
NS67SW 14
Strathclyde Region - Strathkelvin District
Lanarkshire - Cadder ph.

In November 1870 a logboat was discovered (probably during drainage operations) in Cadder Moss, a former loch situated in cultivated clayland near Bishopbriggs and at an altitude of about 60m OD. The stern protruded above the surface and was damaged during the recovery operations.

The boat was found to measure 13' (4m) in length by 1'9" (0.5m) transversely at the bow and 2'5" (0.7m) at the stern, where there was a groove. At least some of the bark was still in place. Two pieces of timber found with the boat had probably been inserted as ribs and secured in place through holes made for that purpose through both the ribs and the boat.

This boat (fig. 19) is in store at Glasgow Art Gallery and Museum under accession number GAGM 01-119. It has been crudely worked from heavily-knotted timber and has suffered from warping and splitting; the surviving part of the starboard side has broken away and is retained in place with iron bands. The underside was inaccessible at the date of visit.

As preserved, the boat measures 3.82m in length by up to 0.58m transversely and the highest surviving point of the starboard side (which has probably not been the sheerline) is 270mm above the floor; the bottom measures about 80mm in thickness and the sides between 30mm and 40mm. The exterior is roughly-worked and rounded but joins the sides at a slight external chamfer; the sectional view of the boat appears incorrect and the displaced side may have been re-mounted at an angle too near the vertical. The form of the bow is a rounded point and stern is square. No trace of the reported transom-groove can be seen and the ribs that were noted when the boat was discovered have since been lost.

The slenderness coefficient and McGrail morphology code for the boat are 5.4 and 1xx:113:323 respectively. The form is dissimilar-ended. *October 1987*.

Dixon 1875.

## 91 Loch Ard
NN 4653 0211
NN40SE 7
Central Region - Stirling District
Perthshire - Aberfoyle ph.

In September 1986 sports divers discovered a possible logboat (which was not recovered) in Loch Ard. It lay in about 5m depth of water at a point 3m S of a crannog (NMRS dataset NN40SE 3) and 370m WSW of the Altskeith Hotel. Loch Ard is one of the lochs that occupy the glaciated valleys of the Trossachs and is at an altitude of about 32m OD.

A section of the boat measuring 3.7m in length, 0.8m in beam and with an interior depth of about 0.5m was identified; the cross-section was noted as semi-circular. One end had been lost but timberwork and heaped stones were thought to indicate the location of the other.

What is apparently this object has been re-located during underwater survey (by Niall Gregory) at a depth of 1.6m and about 5.5m S of the crannog, where the loch-bed is of silt. It was identified as a tree-trunk rather than a logboat, and seen to have split rather than been worked. It is aligned E-W and the exposed portion measures 3.1m in length. The exposed end 'splits into a swallow-tail' at a point about 1.5m from the end, where it measures 0.35m in thickness. No 'heaped stones' were seen in the immediate vicinity. *September 1992*.

*DES*, (1986), 4; NMRS MS/736/4; *pers. comm.* Mr P Dale.

## 92 Loch Arthur 1, and paddle (possible)
NX *c*. 905 687
NX96NW 3
Dumfries and Galloway Region - Nithsdale District
Kirkcudbrightshire - New Abbey ph.

In July 1874 a logboat was revealed near the S shore of Loch Arthur when the water level was abnormally low. This loch (which was formerly known as Loch Lotus or Lotus Loch) is situated on the NW flank of the Criffel massif at an altitude of about 75m OD. Three bronze tri-

*Pl. 10. Loch Arthur 1 (no. 92). View of logboat, apparently soon after its discovery, and said to have been drawn by Professor Geikie. Reproduced from the manuscript collection of the Society of Antiquaries of Scotland held in the Museum of Scotland (SAS Ms. 307/12). I am grateful to the Society of Antiquaries of Scotland for permission to reproduce this plate.*

pod-vessels had previously been found in the loch, and near the N shore there is a crannog. See also no. 93.

Recovery of the logboat was achieved with 'some damage' to the sides and it was subsequently allowed to deteriorate in the open air (pl. 10) while suffering the depredations of visitors. The stern portion (which was the worse affected) was discarded but the forward portion is in store at the Royal Museum of Scotland under accession number NMS IN 3. The short length of a possible paddle-handle that was found in the logboat was also lost; it measured about 2" (51mm) in diameter and the end of it was decorated with carved beading.

The boat was recorded independently by Gillespie and Hewison. It was worked from 'oak' and measured 45' (13.7m) in length over all, and 5' (1.5m) in beam at the stern, which was formed by a transom inserted into a groove about 1½" (38mm) from the extremity. Along the starboard side there survived seven holes measuring about 3" (76mm) in diameter regularly spaced at intervals of about 5' (1.5m). The bottom was pierced by three irregularly-spaced thickness-gauge holes.

The surviving portion of the boat (fig. 18) measures 6.34m in length to the point where it has been sawn off, at which point it measures 0.7m in breadth. The starboard side survives to a height of 0.43m internally and the boat has had an extremely high ratio of length to beam. The remains have not suffered greatly from splitting, but gross warping has displaced the bow section to port. There are possible toolmarks in the exterior of the starboard side and among the numerous knot-holes there are four thickness-gauge holes which are set vertically at irregular intervals close to the rounded junctions of the bottom with the sides. These measure between 35mm by 35mm and 60mm by 40mm and penetrate the bottom which varies between 70mm and 100mm in thickness.

The bow of the boat was recognised to have an unusual elongated form terminating in a 'remarkable prolongation resembling the outstretched neck and head of an animal'. This feature (pls. 11-13) is roughly paralleled at Errol 2 (no. 38) and Loch of Kinnordy (no. 118), and is similarly to be explained as probably fortuitous. The hole that forms the 'eye' measures about 5" (127mm) in diameter and is probably intended to receive a painter or boatrope.

The jagged hole on the port side and the broken section on the starboard are joined by a section of wood which has probably been smoothed, possibly also to hold a rope.

The acutely-angled bow section is also remarkable for the internal 'steps' (pl. 14) that have probably been left as a simple strengthening device. These have no apparent

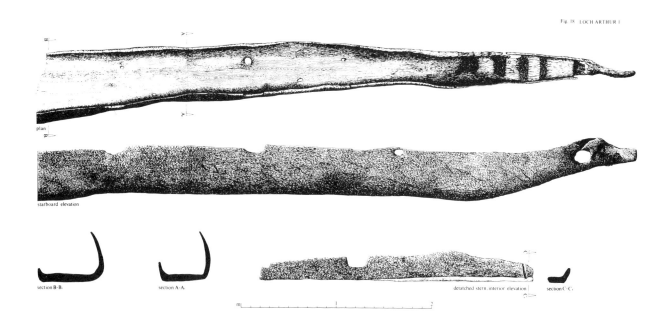

*Fig. 18. Loch Arthur 1 (no. 92). Plan, starboard elevation and two sections of preserved forward section, with interior elevation and (partial) section of detached stern. Scale 1:25.*

Scottish comparanda but are paralleled in three boats from Holme Pierrepont, Nottingham (McGrail 1978, i, 205-12, nos. 57-9), one of which has been dated by radiocarbon to 230 ± 110 bc.

At or near the upper surfaces of the (more complete) starboard side there are two prominent depressions and a hole which measures 90mm by 67mm and is set 30mm below the top of the surviving side at a point where the timber is about 30mm thick. The function of the these features is unclear but they may have served as handholds for portage in spite of the considerable weight of the boat; they were presumably duplicated on the port side.

Stored with this logboat, but having no specific accession number or records, there is a piece of timber (also fig. 18) measuring 2.88m in length which has formed the stern portion of the lower starboard side of a logboat and the adjoining section of the bottom; in the top side there is one deep depression and part of a second. Part of a transom-groove measuring about 25mm in width and 18mm in depth is to be seen and has probably been cut with a metal chisel. Although the timber of this fragment is finer-grained and more heavily knotted than the main section of the boat, the presence of the depressions and of the transom groove make it probable that this fragment has formed part of the same boat.

On the basis of the best available measurements, the slenderness coefficient of this vessel was 9, the beam/draught coefficient was 3 and the displacement under standard conditions about 18.8 cubic metres. This was without doubt a very large logboat by Scottish standards, having speed potential, manoeuvrability and ample capacity for bulky cargo. The McGrail morphology code is 44a3:3x4:331 and the form dissimilar-ended.

This logboat has been radiocarbon-dated to 101 ± 80 bc (SRR-403) although the location of the sample makes this date erroneously early. The determination may be calibrated to about 75 cal BC, and the vessel should probably be dated within the last century and a half BC and the first two centuries AD. *August 1987.*

*PSAS*, xi (1874-6), 19-20; Gillespie 1876; Hewison 1939, 47-8; Close-Brooks 1975; McGrail 1987a, 82, 84.

93 **Loch Arthur 2**
NX 9028 6898
NX96NW 1
Dumfries and Galloway Region - Nithsdale District
Kirkcudbrightshire - New Abbey ph.

In 1966-7 archaeological survey revealed a 'possible fragment of a dug-out canoe' on the SE side of the crannog in Loch Arthur and about 15m SSW of the remains of a possible jetty. This object was apparently left *in situ* but was not re-located in August 1992 during further underwater survey (by Niall Gregory).

Williams 1971, 123; NMRS MS/736/4.

**Loch Awe** See **Carn an Roin**

II  *Gazetteer of Logboat Discoveries*

*Pl. 11. Loch Arthur 1 (no. 92).*
*Views and details reproduced by kind permission of the Trustees of the National Museum of Scotland:*
*view of port side, from bow.*

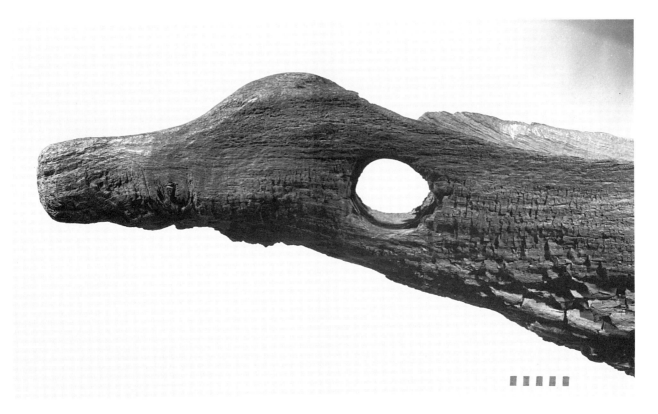

*Pl. 12. Loch Arthur 1 (no. 92).*
*Views and details reproduced by kind permission of the Trustees of the National Museum of Scotland:*
*detail of port side of 'animal-head' bow.*

*Pl. 13. Loch Arthur 1 (no. 92). Views and details reproduced by kind permission of the Trustees of the National Museum of Scotland: detail of starboard side of 'animal-head' bow.*

*Pl. 14. Loch Arthur 1 (no. 92). Views and details reproduced by kind permission of the Trustees of the National Museum of Scotland: detail of stepped interior of forward section, looking forward.*

**Loch Canmor** See **Loch Kinord**

**94  Loch Chaluim Chille 1**
NG c. 37 68
NG36NE  8
Highland Region - Skye and Lochalsh District
Inverness-shire - Kilmuir ph.

In 1763 a logboat was revealed during the drainage of Loch Chalum Chille or St Columba's Loch which was formerly situated in a depression about 1km from the sea in the Trotternish district of Skye, and at an altitude of about 15m OD.

The boat was discovered 'deeply imbedded in the bottom of the lake'. It measured about 14' (4.3m) in length by 3' (0.9m) in beam and was 'much stronger and far more firmly built than any of modern date'. The timber was variously identified as 'oak' or 'fir'. Fastened to the ends were five iron rings 'of almost incredible thickness' which were reworked into agricultural implements. The boat itself was destroyed.

*NSA*, xiv (Inverness), 267; Jolly 1876, 553-5, 559-61.

**95  Loch Chaluim Chille 2**
NG c. 373 688
NG36NE  14
Highland Region - Skye and Lochalsh District
Inverness-shire - Kilmuir ph.

In 1874 a second logboat was revealed during the cleaning of the deep ditch that runs along the W side of the former loch. It lay horizontally at a depth of about 4' (1.2m) and the stern was removed by the finders. Unsuccessful attempts were made in the following year to recover the remainder of the boat.

Jolly recorded the stern as being rounded and measuring about 6" (150mm) in height; the bottom was split and within the interior he noted toolmarks but no evidence for the use of fire. He also noted grooves along 'the middle of the thickness of the stern' and along 'the outside of the bottom, running from stem to stern in all likelihood', but the function of these features is unclear. The timber was variously suggested to be alder, fir, pine or oak, and was in poor condition.

Jolly 1876, 555-61.

**Loch Davan**  See **Loch Kinord**

**Loch Dernaglaur**  See **Dernaglar Loch**

**96-101  Loch Doon 1-6, and paddle (possible)**
NX c. 488 947
NX49SE  11
Strathclyde Region - Cumnock and Doon Valley District
Ayrshire - Straiton ph.

In the early nineteenth century logboats were found near the former island site (at NX 4882 9475) of Loch Doon Castle (NMRS dataset NX49SE 1) with which they have traditionally been associated. Some at least of them were recovered from among 'a great many canoes...lying in all directions' which may have been the timbers of a structure ancillary to the castle or, possibly, a crannog.

In about 1935 the castle was re-erected on the W shore (at NX 4847 9499) and the water level raised to convert the loch (which is situated in a glaciated valley at an altitude of about 215m OD) into a major reservoir.

There were also found 'in or near' the logboats a wide variety of objects including a medieval battle-axe, a mophead (which, it was suggested, had possibly been used for spreading pitch on the logboats) and a 'rude oak club' (possibly a paddle) measuring about 3' (0.9m) in length.

Most accounts follow the *New Statistical Account* in assigning three boats each to the years 1823 and 1831 but the more detailed article by Cathcart mentions four discoveries in 1831 and two 'some years since', presumably in 1823.

1. One of the logboats that was found in 1823 was taken to the 'Museum of Glasgow University' (now the Hunterian Museum, University of Glasgow). Although it was not recorded in detail upon its discovery, it can be identified with that displayed under accession number HM A30.

After shrinkage, this very simple boat (fig. 20) measures 3.4m in length over all, up to 0.86m in beam and up to 0.41m in external depth. The bow is of barge-like form and is rounded both internally and externally, while the stern is formed by a vertically-cut section from which a pronounced horizontal shelf of duck-bill form projects rearwards. Evidence of charring and the marks of a probable metal tool measuring about 35mm in breadth are visible in the interior, and the chamfer along the lower part of the sides has been cut with an adze. The bottom of the boat is noticeably uneven, giving rise to a suggestion that it remains unfinished; in the absence of thickness-gauge holes or major splits its thickness cannot be accurately determined but it is possible that the uneven bottom results from an uncomprehending attempt to construct a logboat by a process of trial-and-error without the usual aids to stability control.

The boat has been worked from a nearly-whole section of log and has a slenderness coefficient of 4, a beam/draught coefficient of 2.1 and a displacement of 0.7 cubic metres; these figures which denote a small boat with a relatively high capability for carrying bulky cargoes. The form is dissimilar-ended and the McGrail morphology code is 111:111:521.

This boat has been dated by radiocarbon to 509 ± 110

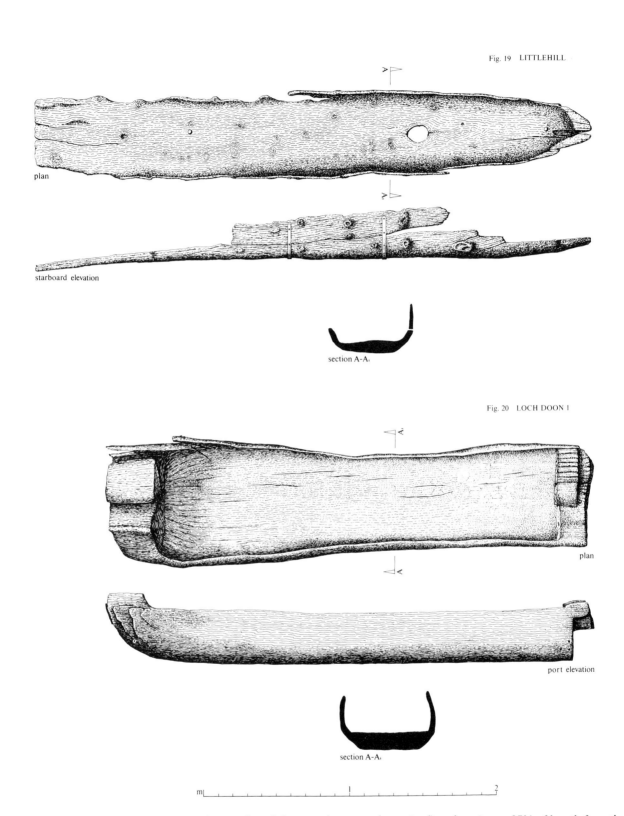

*Fig. 19. Littlehill (no. 90). Logboat. Plan, starboard elevation (as currently retained) and section at 37% of length from the bow. Scale 1:25.*

*Fig. 20. Loch Doon 1 (no. 96). Logboat. Plan, port elevation and section amidships. Scale 1:25.*

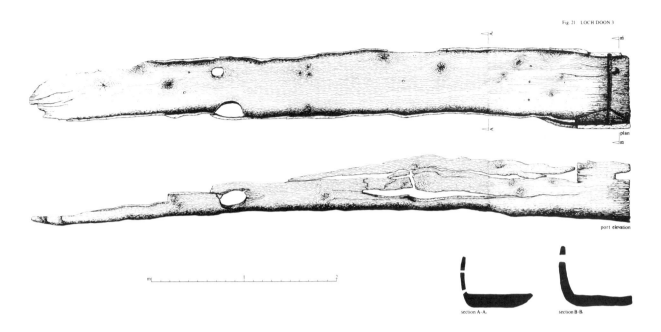

*Fig. 21. Loch Doon 3 (no. 98). Logboat. Plan and port elevation with sections at 77% of length from bow, and at stern. Scale 1:25.*

ad (SRR-501), which determination may be calibrated to about 619 cal AD. The sample has, however, been taken from a point where a considerable volume of timber has been removed, and the date obtained is possibly about a half-century older than the true felling date.

The experimental reconstruction (in a finished form) of this vessel by students of the Scottish Institute of Maritime Studies, University of St Andrews (under the supervision of Damian Goodburn) and the subsequent display of the replica at the Scottish Fisheries Museum, Anstruther, Fife serve both to provide an interesting display item and give a vivid impression of the practical constraints upon the manufacture and use of logboats in general. *August 1987.*

2. The second boat found in 1823 was not recorded in detail, and is lost.

3. Each of the three larger logboats that were found in 1831 was extracted with great difficulty from 'nearly a vertical position' and found to measure about 23' (7m) in length over all, 2'5" (0.7m) in depth externally and 3'9" (1.2m) in beam internally across the stern which was closed by a board. Each was of 'oak', free of ornament, and had been covered with pitch, both internally and externally. One of them was initially removed to a pond at Berbeth (NS 46 03) and subsequently (probably about 1917) taken to the Hunterian Museum where it is displayed under accession number HM A31.

This boat (fig. 21) now measures 6.55m in length and up to 0.85m in beam, being shaped as a long triangle and broadest near the stern. It has suffered both warping and splitting and has been glued in places, presumably during conservation. The timber contains numerous knots (some of them of considerable size) as well as seven thickness-gauge holes which measure up to 30mm in diameter; four of these retain their plugs. The flat bottom measures about 140mm in thickness. The sides have risen at near right-angles; the port side measures about 50mm in thickness and survives to a maximum height of about 0.5m, but the starboard side is almost entirely missing. Slight traces of what may be toolmarks are to be seen on the exterior of the port side amidships.

The form of the bow is now difficult to ascertain as most of it has been lost, but it was most probably rounded in all three planes. The stern was formed by a transom, the groove for which measures 50mm in breadth and about the same in depth. The lower part of the transom survives *in situ* and measures 630mm transversely across the boat, 90mm in height to a split flat top, and about 30mm in thickness.

Several features that were noted in the original accounts of the discovery cannot now be identified, probably on account of the loss of the upper parts of the sides. In the bow there was a round hole, possibly to hold a tow- or boat-rope, and there were two grooves across the boat; that located amidships was probably intended to receive the ends of a thwart. The transom-groove was said to supplemented by 'two strong pins of wood passing through well cut square holes, at a little distance from the end on each side'.

On the basis of the measurements recorded at the time of discovery and on the assumption that the sides do not survive to their full height, the slenderness coefficient is 8.1, indicating a considerable speed potential. Assuming the form of the bow to be as postulated, the McGrail morphology code for this boat is 441:2x1:222 and the form dissimilar-ended. *August 1987.*

4-5. The other two logboats that were found in 1831 were of approximately the same dimensions as (3). They were 'removed...in mere fragments', which are now lost.

6. The smallest of the 1831 discoveries measured 12' (3.7m) in length over all, 2'9" (0.9m) in beam at the stern and 2' (0.6m) in depth externally. Both bow and stern were of squared form, the former being undercut. This boat was removed to Berbeth, but is now lost.

On the basis of these measurements, the slenderness and beam/draught coefficients may be estimated at 4.4 and 1.4 respectively. The displacement was of the order of 1.1 cubic metres and the McGrail morphology code was apparently 1xx:xxx:12x while the form was dissimilar-ended.

NSA, v (Ayr), 337-8; Cathcart 1857; Mackie 1984; *pers. comm.*, Mr. D. Goodburn.


**Loch Dowalton**  See **Dowalton Loch**


**102 Loch Glashan 1**
NR *c.* 920 934
NR99SW 11
Strathclyde Region - Argyll and Bute District
Argyll - Kilmichael Glassary ph.

In 1960 a logboat was discovered (pls. 15-17) on the E side of Loch Glashan when the level of the loch was lowered during the construction of a dam. This inland loch is situated at an altitude of about 100m OD near the NW shore of Loch Fyne.

The boat measured 11' (3.4m) in length and had a transom stern. A midships thwart rested on blocks left in the solid and a stempost worked in the solid was seen to continue beneath the boat as a solid false keel. A 'fragment of twisted rope' was found in the hole through the bow, but does not appear to survive, while the 'possible wooden paddle or oar' that was found close to the logboat, and was possibly associated with it, is described below (no. A36).

In the following year the boat (fig. 22) was removed to Glasgow Art Gallery and Museum where it is stored under accession number GAGM A6137a; it is mounted in a cradle and the port side was inaccessible at the date of inspection. Severe flaking and partial disintegration have occurred on the lower part of the exterior in spite of conservation with Carbowax. Splitting is especially evident in the starboard side and at the ends, where warping has also caused gross distortion. There is one knot of considerable size in the bottom.

After shrinkage, the boat measures 3.06m in length over all by up to 0.82m in beam (at a point near the stern) and up to 0.78m in external depth. Much of this latter measurement is accounted for by the raised and distorted bow section, and for the greater part of their length the sides only survive to an external height of about 0.3m.

The boat has been crudely shaped without the use of thickness-gauges, and the interior is rounded in section. The bottom measures between 80mm and 100mm in thickness, and the sides about 40mm. Two lengths of timber have been left in the solid near the bow; that on the port side measures 590mm in length and has its upper surface about 60mm above the floor while the corresponding figures for the starboard side are 430mm and 30mm respectively. Resting on these two projections there is a loose-fitting thwart which measures 610mm by 120mm and 55m in thickness and is squared in all planes except under the lower corners where it is rounded to conform with the sides.

The bow is rounded in form, although heavily distorted by warping, and bears heavy radial splitting internally. It is pierced transversely by a horizontally-bored hole which measures about 20mm in diameter and 90mm in depth; a length of what was probably a boat-rope was found in this hole but has since been lost. The prominent external cutwater is squared in form and measures about 30mm along the centreline by 50mm transversely; this extends around the forefoot and along the bottom of the boat to terminate at the stern with a downwards flare to a maximum depth of 35mm. The stern itself is square in form although it appears to have been displaced at an angle by warping. The transom is now lost, but its groove extends to the highest surviving point of the side and measures 20mm in width and 25mm in depth. The sharpness of the edges indicates that it was cut with a metal tool, probably a chisel.

The boat has apparently been worked from a whole log which measured about 0.8m in diameter. On the assumption that the sides extended to the full height of the present bow section, the conversion coefficient was about 81%, this low figure reflecting the thickness of the sides and bottom and the size of the block of timber that forms the bow section. On the basis of the same assumption, the slenderness and beam/draught coefficients were 3.1 and about 1.06 respectively, and the displacement was about 1.15 cubic metres, indicating that this boat was beamy, directionally stable and best-suited to the carriage of such high-density cargoes as men and animals. The block and volumetric coefficients are 0.59 and 0.04 respectively and the McGrail morphology code is 14a2:2x2:222 while the form is dissimilar-ended. *October 1987.*

DES, (1960), 6; Campbell and Sandeman 1962, 120, no. 56b; McGrail 1987a, 80, 84; RCAHMS 1988, 205-8, no. 354.


**103 Loch Glashan 2**
NR *c.* 916 925
NR99SW 2
Strathclyde Region - Argyll and Bute District
Argyll - Kilmichael Glassary ph.

*Pl. 15. Loch Glashan 1 (no. 102). Photograph taken at the time of discovery and reproduced from the negatives held in the NMRS (A53503-5) by kind permission of Mr. J.G. Scott and of the Royal Commission on the Ancient and Historical Monuments of Scotland: general view from the port side.*

*Pl. 16. Loch Glashan 1 (no. 102). Photograph taken at the time of discovery and reproduced from the negatives held in the NMRS (A53503-5) by kind permission of Mr. J.G. Scott and of the Royal Commission on the Ancient and Historical Monuments of Scotland: view from the port bow, showing the general form of the boat, the transom and the thwart.*

*Pl. 17. Loch Glashan 1 (no. 102). Photograph taken at the time of discovery and reproduced from the negatives held in the NMRS (A53503-5) by kind permission of Mr. J.G. Scott and of the Royal Commission on the Ancient and Historical Monuments of Scotland: view from astern..*

In 1960-1 a crannog (at NR 9159 9249) and a medieval island-settlement (at NR 9168 9254) were excavated in advance of engineering works. During the investigation of the latter feature 'fragments of a dug-out canoe' were noticed 'at the northern end of the strait' that separated the monument from the shore. The OS locate the discovery at NR 9173 9252, but the annotated photograph that accompanies the published report indicates a location adjacent to the settlement. The vessel may remain *in situ*.

Fairhurst 1969, 47; RCAHMS 1988, 205-8, no. 354.

**104 Loch Kinellan**
NH 4710 5759
NH45NE 7
Highland Region - Ross and Cromarty District
Ross and Cromarty - Contin ph.

In 1914-17 HA Fraser, assisted by Father Odo Blundell and Robert Munro, excavated the crannog in Loch Kinellan, which is situated in upper Strath Peffer at an altitude of about 125m OD. The crannog had previously been identified as the hunting-seat of the Earls of Ross in the fourteenth and fifteenth centuries and pottery of this period was discovered.

The excavations comprised a number of small trial-pits and in 1915 the excavation of pit no. 6 revealed a logboat in the centre of the substructure at a depth of about 8' (2.4m) below the highest point. A 'large number of bones' were found near the boat and immediately above it there was a 'very fine' flint flake. The pit was subsequently enlarged and the boat was extracted during the next season. It was taken to the (former) museum at Fort Augustus Abbey but cannot be located and is said to have 'disintegrated on being exposed to the air'.

When discovered, the incomplete boat measured 24'9" (7.5m) in length and 'probably' 2'6" (0.8m) in beam; it had probably been 'considerably damaged' before it suffered considerable warping through use as a crannog timber. The incomplete nature of the recorded remains precludes quantitative analysis or assessment of form.

Fraser 1917, 50, 87-9, 94, figs. 15 (xix) and 22; *pers. comm.* Mr. J.A. Grieve.

**105-8 Loch Kinord 1-4**
Grampian Region - Kincardine and Deeside District
Aberdeenshire - Glenmuick, Tullich and Glengairn ph.

The discovery of four logboats is recorded from Loch Kinord, which has also been known as Loch Kinnord and Loch Canmor and is situated on the NE fringe of the

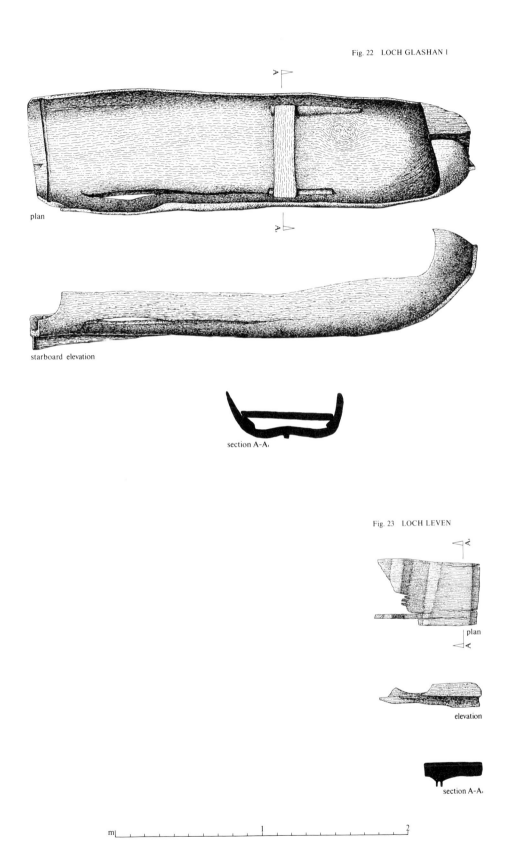

*Fig. 22. Loch Glashan (no. 102). Logboat. Plan, starboard elevation and section midships. Scale 1:25.*
*Fig. 23. Loch Leven (no. 116). Plan, elevation and surviving remains of possible logboat. Scale 1:25.*

Grampian massif, at an altitude of about 165m OD. A crannog (NMRS NO49NW 17), an island-castle (NMRS NO49NW 16) and various items of medieval and post-medieval pottery and metalwork have also been found in the loch. See also nos. A45-6.

These may be the 'Canoes' that have been recorded (apparently in error) in the nearby Loch Davan (NJ c. 44 00).

1. NO c. 443 996   NO49NW 30

The logboat that subsequently became known as the 'Royal Yacht' was discovered 'near the north shore and opposite the smaller island', presumably the crannog at NO 4433 9952. It had been revealed during several periods of low water before 1858, when its discovery was decided upon; this operation was carried out in June of the following year. The boat was taken to Aboyne Castle where it appears to have disintegrated; what is probably an unplaceable fragment of it remains in store at Aberdeen University Anthropological Museum under accession number AUAM 14900, having been received through the Ogston Collection.

On recovery, the boat was found to measure 22'6" (6.9m) in length and 3'3" (1m) in breadth at the squared stern; the bow was pointed. The sides survived to a maximum height of 9" (225mm) and the floor was about 4" (100mm) thick. What were probably five pairs of thickness-gauge holes were seen to pierce the bottom; each hole measured about 1¾" (45mm) in diameter, and they were set at irregular intervals of between 1'6" (0.4m) and 1'9" (0.6m) between the stern and a point 6'6" (2m) from the prow. The slenderness coefficient was about 6.9.

Splitting had apparently taken place while the boat was still in use and there was evidence of an attempted repair. Two 'bars of oak' had been countersunk longitudinally into the underside of the boat and were held in place by five transverse bars, each of which measured between 1'10" (0.6m) and 2'3" (0.7m) in length and 3" (75mm) in width. These transverse bars were similarly countersunk into the boat; they were dovetail-jointed into the longitudinals, and were held in place by a 'wooden bolt' at each end.

2. NO c. 437 996   NO49NW 31

In August 1875 (and again during a period of lowered water level) Lord and Lady Huntly recovered a logboat which had been noticed previously 'about 80 yards from the north shore and 200 yards due west from the Castle Island'. The castle is at NO 4397 9964 and the logboat was probably found around NO 437 996.

On recovery, the boat was found to measure 30'2" (9.2m) in length, and between 3'5" (1m) and 3'7" (1.1m) in beam. The timber was identified as 'oak'. The sides did not survive to a significant height and the bottom had been worn down to a thickness of about 7" (175mm). Across the bottom there were four 'ribs or ridges...at nearly equal distances from each other' which were identified as foot-rests for the oarsmen but were more probably strengthening-ribs left in the solid. The slenderness coefficient was 8.4, but the form cannot be identified on the basis of the published account.

The boat was housed by Sir W.C. Brook in a specially-constructed shed, but no remains of it can now be traced.

3. NO c. 437 996   NO49NW 31

A further logboat was revealed and brought to land during the recovery of (2) and at the same location; it was similarly placed in storage by Sir WC Brooks, and its fate is likewise unrecorded.

This boat measured 29'3" (8.9m) in length and was of slightly greater beam than that found previously; the slenderness coefficient may accordingly be calculated at approximately 7.5. The general condition of the boat was similar to that found previously and probable strengthening-ribs were similarly noted.

4. NO c. 4433 9952   NO49NW 32

In 1875 a boating party discovered a fourth logboat about 30 yards (27.5m) SE of the Prison Island crannog. It was apparently filled with stones.

The boat was not brought to shore until 1962 when it was re-discovered by sports divers. The remains were found to measure 12' (3.7m) in length and up to 2'4" (0.7m) transversely. The timber was identified as 'oak'. What was probably the stern was pierced by a 'worn' hole which measured about 3" (76mm) in diameter. After measurement the boat was returned to the loch. On the basis of the incomplete record available, the slenderness coefficient may be calculated at about 5.1.

In August 1992, underwater survey (by Niall Gregory) failed to re-locate these remains in extremely silty bottom conditions, either near the location stated above or at an alternative location (suggested on the basis of hearsay by the Warden of the National Park) between the crannog and the island-castle at NO *c.* 442 995.

Hogg 1890, 160; Michie 1910, 84-6, 88-9; Ogston 1931, 32-3; *DES*, (1962), 1; *Quaternary Research Association, Field Excursion Guide, April 1975*; NMRS MS/736/4; *pers. comm.* Dr. K.J. Edwards.

109-15 **Loch Laggan**
Highland Region - Badenoch and Strathspey District
Inverness-shire - Laggan ph.

Seven logboats and the remains of a framed boat of unknown date (no. A48) have been discovered during periods of low water level on the shores of Loch Laggan, an extensive highland loch which occupies a deep glaciated valley in the catchment area of the River Spean at an altitude of about 280m OD and is used for the generation of hydro-electric power. The various accounts of the five earlier discoveries were collated in 1951 but those made subsequently have received only brief publication.

1.  NN 4987 8755   NN48NE 1

In 1934 a logboat of 'fir' was revealed on Eilean an Righ or King Fergus' Isle, a medieval island-dwelling which was traditionally a hunting-lodge of the kings of Scotland. The boat was 'in two parts and very frail'; it has apparently not survived.

2.  NN c. 535 895   NN58NW 6

It was also in 1934 that a second 'fir' logboat was found at the extreme E end of the loch, near the mouth of the River Pattack; an incomplete wooden bowl (no. A47) was found in it. On the basis of the detailed contemporary record that has survived, this boat is probably the example that Maxwell illustrates (pl. 18); this photograph was probably taken at the time of initial discovery as it is attributed to J.M. Corrie who died in 1938. The OS are apparently in error in suggesting a location on the N shore of the loch and probably near the W end.

In 1949 the boat was rediscovered and found to measure 37'7" (11.6m) in length.

It is said to have been initially found in peaty silt at a point about 60' (18.3m) from the old shoreline and where the water would have been about 5' (1.5m) deep; the timber was generally about 2" (51mm) thick and the boat appeared 'remarkably well preserved'. No evidence of metal fittings or the use of metal tools was noted.

Several unusual features of this boat make its apparent loss a matter for regret. At the bow there was a feature which has been described as a 'figure head' but appears from the photographs to have been a fortuitous projection sited away from the centreline of the boat.

Along the bottom of the mid-section (and to port of the centreline) there are two longitudinal battens which were secured by substantial lashings or (less probably) cleats. These were probably inserted to plug a rapidly-developing split and may be identified with the 'patch...applied by its prehistoric owners' that is noted in the published account. The repair was described at the time of its initial discovery as 'a plank some four inches (102mm) wide and perhaps one inch (25mm) thick held firmly in position by rough cross pieces that fitted so snugly over and around it into holes bored into the floor that they might have come from some very pliable material such as tree root'. This is probably to be interpreted as an attempt to close the split by pulling the sides together, oversewing a batten which served to retain the caulking material.

At the stern there was a probable transom pierced by what was identified as a 'socket' for a steering-oar but was more probably a knot-hole, as the insertion of an oar of any length would have been inconvenient and such an orifice would be a source of severe leakage unless protected by flexible sleeving. Finally, lying 'at about the same position' as the repair there was a 'roughly fashioned piece of wood that looks like a sapling' and was possibly a punt-pole.

On the evidence of the available photographs, the McGrail morphology code of the boat was 4x2:112:322. The boat was of narrow and dissimilar-ended form, but this cannot be quantified.

3.  NN c. 533 889   NN58NW 3

In August 1948 the remains of a 'fir' logboat measuring about 30' (9.2m) in length was seen 'in the sand' at the E end of the loch, and to the S of the mouth of the River Pattack. It was not recorded in detail.

Although this was possibly a re-discovery of (2), the recorded locations are sufficiently far apart to justify the separate recording of the two discoveries. The OS suggest that this boat survives underwater at NN 5335 8894 and is exposed intermittently. In August 1992, beach-walking and underwater survey (by Niall Gregory) failed to locate these remains.

4.  NN c. 499 873   NN48NE 2

In 1949 a further 'fir' logboat was discovered on the S shore of the loch, just E of King Fergus' Isle. Nothing more is known of this discovery and it was possibly a re-exposure of (1). In August 1992, beach-walking and underwater survey (by Niall Gregory) failed to locate these remains.

5.  NN c. 520 891   NN58NW 4

Also in 1949, BM Peach noted the 'undoubted remains of a canoe bottom' in a bay on the N side of the loch, about '1¼ miles (2km) from the eastern end'. The national grid reference was ascertained as NN 520 890 at the time of discovery, but re-plotting on a map of larger scale suggests a slight revision to the position close inshore at Tullochroam that is cited above.

The remains measured 14'6" (4.4m) in length by between 1'5" (0.4m) and 2'6" (0.8m) in beam. The timber measured between 1½" (40mm) and 2" (50mm) thick and was identified as 'oak'. Its condition was 'much too tender' to withstand removal and the boat probably exists no longer. On the basis of the recorded measurements, and assuming the full length of the boat to have been preserved, the slenderness coefficient was 5.8. The form is not recorded.

In August 1992, beach-walking and underwater survey (by Niall Gregory) failed to re-locate these remains.

6 and 7.  NN c. 426 849   NN48SE 1

In 1955 two 'partly buried' logboats were found on the N shore of the loch. It is probable that neither of these boats survives.

*Pl. 18. Loch Laggan 2 (no. 110). General view from the bow, showing evidence of repair. Reproduced from plate xvi (lower) of volume lxxxv (for 1950-1) of the* Proceedings of the Society of Antiquaries of Scotland. *I am grateful to the Society of Antiquaries of Scotland for permission to reproduce this plate.*

6. One of them was brought ashore and found to measure 17'6" (5.3m) in length and between 1' (0.3m) and 1'8" (0.5m) in breadth, being broader at the bow. Along the bottom there were two rows of what were probably plugged thickness-gauge holes, and at the stern there was a 'square slot' which was said to be for a 'steering oar or tiller'. The form of the boat cannot be determined, but on the basis of the recorded measurements the slenderness coefficient was unusually high at 10.5; it is possible that the boat had split and that full beam was not recovered.

Maxwell 1951, 163-5; *DES*, (1955), 17; McGrail 1978, i, 11, 37; McGrail 1987a, 65; NMRS cancelled dataset NN48SE 2; NMRS 1/6 (letter dated 1 Dec 1988); NMRS MS/736/4.

**116 Loch Leven**
NO *c*. 13 01
NO10SW 13
Tayside Region - Perth and Kinross District
Kinross-shire - Kinross ph.

Burns-Begg mentions that a logboat was revealed when the level of Loch Leven was lowered by drainage operations. It was taken to Kinross House and is no longer to be seen. The date of discovery is unclear from his accounts but it was probably within the period 1830-2 when the level of the loch is said to have fallen by about 4'6" (1.4m) following the construction of the New Gullet drainage canal.

In Perth Museum and Art Gallery there are two timber fragments which have possibly formed part of this boat. One of them (accession number PMAG K1972.42) is roughly pyramidal in shape and measures about 130mm square at the base by about 90mm in height. It has suffered badly from splitting and its origin cannot now be determined but its thickness suggests that it may have formed part of the bow. The other fragment (accession number PMAG K1972.286) measures up to 360mm by 230mm by 80mm and may be identified (fig. 23) as the badly-split remains of the aftermost part of the floor of a logboat. There is a pronounced round-down at one end and the worn remains of a transom-groove are recognisable; this measures about 120mm in breadth at the top, 90mm in breadth at the bottom, and about 30mm in depth. *August 1987.*

Burns-Begg 1887, 4; Begg 1888, 118; Burns-Begg 1901, 5; Walker 1980, 29.

**Loch Lotus** See **Loch Arthur**

**117 Loch nam Miol**
NM *c*. 518 527
NM55SW 3
Strathclyde Region - Argyll and Bute District
Argyll - Kilninian and Kilmore ph.

In about 1870 a 'large' logboat, several 'canoes of a smaller size' and three boats of clinker construction were discovered during the drainage of Loch nam Miol, which has also been known as Loch na Mial or Lochnameal and formerly occupied a depression in the craggy terrain of the Aros district of NE Mull at an altitude of about 80m OD.

The largest logboat was found at a depth of 4' (1.2m) close to the causeway leading to the crannog in the loch (at NM 5185 5273) and measured 17' (5.2m) in length by 3'6" (1.1m) in beam. The timber was identified as 'black oak' and was 'quite fresh and sound'. An unsuccessful attempt was made to preserve the boat by immersion in seawater.

On the evidence of the recorded measurements the slenderness coefficient of the boat was unusually small at 4.8. This may indicate that the full length was not preserved. The shape of the boat was not noted.

Campbell 1870; Duns 1883, 350; RCAHMS 1980, 122, no. 251.

**118 Loch of Kinnordy**
NO *c*. 36 54
NO35SE 12
Tayside Region - Angus District
Angus - Kirriemuir ph.

In 1820 a logboat (pl. 19, central feature) was discovered during marl-digging in the partly-drained Loch of Kinnordy, which is situated in cultivated clayland at an altitude of about 140m OD. It was near the crannog that had been revealed during eighteenth-century drainage operations and has been identified with a mound at NO 3667 5441. The published section (pl. 19, lower band) shows the boat to have lain horizontally in the upper layer of marl and immediately below the peat layer. Hutcheson adds that it was found at a depth of about 10' (3m) and that a 'thunderbolt axe' (probably a polished stone axe) was found in it but subsequently lost. Cervid bones were also found in the layer.

The boat was recorded by (Sir) Charles Lyell, the eminent geologist, and was found to be of 'oak' it measured 15' (4.6m) in length 'from bow to stern' and its 'width within' was 3' (0.9m). No evidence of propulsion was seen but within the stern there was a 'seat for the boatman' while the prow 'had evidently been carved into an ornamental shape, representing, apparently, the head of some animal'. The boat was analysed in its geological context and assumed to be of prehistoric date, as Lyell noted that 'it shows, that some part even of the peat, which overlies all the marl, is of a date anterior to the historical records of the county'. The boat has, however, since been radiocarbon-dated to 735 ± 40 ad (Q-3142), which determina-

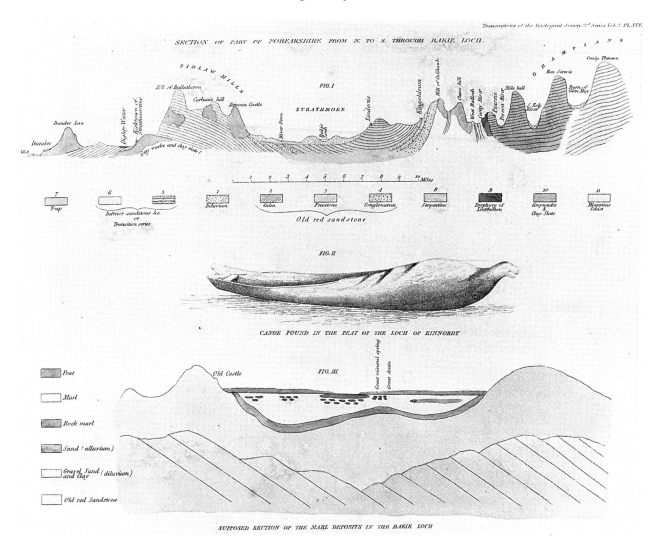

*Pl. 19. Loch of Kinnordy (no. 118). Publication drawing by Lyell, showing the stylised depiction of the suggested animal-head and the setting of the boat in its geological context. Reproduced, by kind permission of the Geological Society and of Edinburgh University Library, from the* Transactions of the Geological Society of London *for 1829.*

tion may be calibrated to about 791 or 801 cal AD.

The boat (fig. 24) was kept for some years in Kinnordy House but is now in store at Dundee Museum and Art Gallery under accession number DMAG 69-256. The starboard side was difficult of access at the date of visit. It has suffered considerable splitting and gross deformation through warping so that the bow has been twisted through about 45° to port, causing a much-distorted plan view to be presented. The large gash in the forward section of the starboard side was possibly caused during recovery operations.

The workmanship displayed in the boat is of an unusually high order, there being little differentiation internally between the sides and the bottom, which only measure between 20mm and 40mm in thickness in spite of thickness-gauges not having been used. The timber is generally smooth and relatively free of knots, and there are slight traces of possible toolmarks internally amidships and near the stern. The exterior displays a slight chamfer in the lower part of the sides.

After shrinkage, the boat measures 4.32m in length over all and up to about 0.85m in beam. It has probably been parallel-sided on plan although the variation in the heights of the side presents a misleading impression in plan view. The sides themselves survive to a height of about 0.35 amidships but rise to about 0.65m at the pointed bow, where the possible animal head that Lyell noted is probably to be explained as fortuitous, although the hole may have served to retain a boat-rope. The stern is solid and square-cut internally but is externally sub-rectangular. The fitted seat has been lost and no support for it can now be distinguished.

Making allowance for distortion in shrinkage, the slenderness coefficient of this boat was 4.3, the beam/draught coefficient was 1.3 and the displacement under standard conditions was about 1.43 cubic metres, indicating that this was a small and general-purpose craft. The McGrail morphology code for this boat is 532:1x2:352 and the form is dissimilar-ended. *August 1987.*

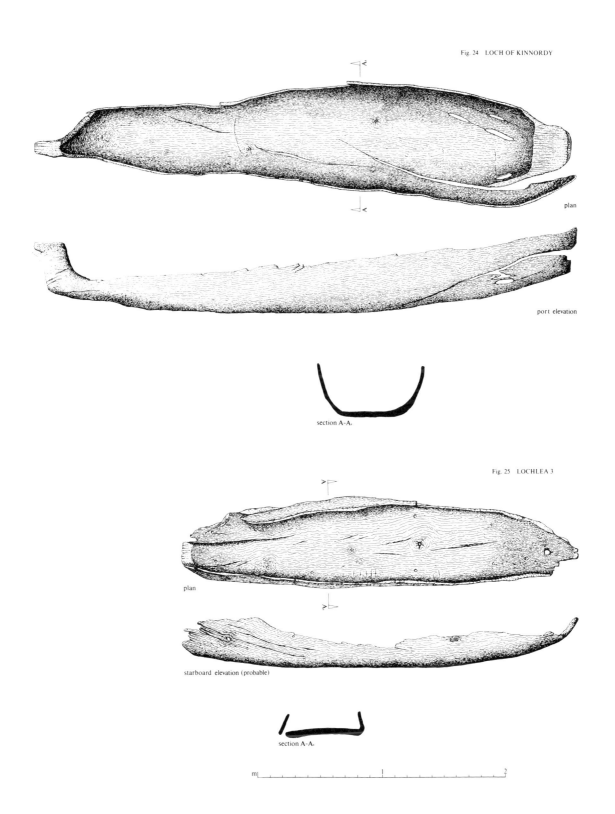

*Fig. 24. Loch of Kinnordy (no. 118). Plan and port elevation of surviving remains, with section amidships. Scale 1:25.*
*Fig. 25. Lochlea 3 (no. 126). Plan, elevation of (probable) starboard side and section amidships. Scale 1:25.*

Lyell 1829, 87-8 and accompanying plate; Warden 1880-5, iv, 115; *[Dundee] Evening Telegraph*, 13 August 1895; *[Dundee] Evening Telegraph*, 19 August 1895; Hutcheson 1897, 270-1; Coutts 1971, 67; McGrail 1978, i, 67; McGrail 1987a, 84; *pers. comm.* Miss C. Lavell.

**119-20  Loch of Leys 1-2**
NO *c.* 70 97
NO79NW 2
Grampian Region - Kincardine and Deeside District
Kincardineshire - Banchory-Ternan ph.

In 1850 two probable logboats were discovered 'at some distance apart' during drainage operations in the Loch of Leys (which has also been known as the Loch of Banchory). One of them was located at NO 7051 9790, about 100m NNE of a crannog. Their fate was not recorded.

1. One of the boats measured about 9' (2.8m) in length. Two nails were used in the construction while the bottom was 'flat and composed of one piece of oak'.

2. The other boat was 'small'.

Six bronze vessels may have been found under one of these logboats, which was inverted. Two of these are in the Royal Museum of Scotland (under accession numbers NMS MA 6-7) and the others are at Crathes Castle.

*PSAS*, i, (1851-4), 27; Name Book, Kincardineshire, No. 3, pp. 77-8; OS 6-inch map, Kincardineshire, 1st ed. (1869), sheet v.

**Loch of Sanquhar**  See **Black Loch**

**121  Loch of the Clans**
NH 8260 5298
NH85SW 1 and 4
Highland Region - Nairn District
Nairnshire - Croy and Dalcross ph.

About 1823 the 'remains of a Canoe' were revealed during drainage operations within the area formerly occupied by the Loch of the Clans, and now noted as Muir of the Clans. The *New Statistical Account* records that the boat was 'of most beautiful workmanship' but was 'cut down for mean and servile purposes by some modern Goth'.
  The Ordnance Survey locate the discovery at a point 170m SW of Bemuchyle farmhouse and 30m ENE of a crannog in an area of prominent eskers running from ENE to WSW at an altitude of about 30m OD.
  This is probably the logboat that Wallace notes at 'Nairnside'; the modern farm of that name is situated about 2.7km to the S.

*NSA*, xiv (Nairn), 448; Name Book, Nairn, No. 5, p. 24; *PSAS*, v (1862-4), 117; *Transactions of the Inverness Scientific Society and Field Club*, ix (1918-25), 130.

**Loch Sanish**  See **Parkfergus**

**Loch Shiel**  See **Acharacle**

**Loch Tay**  See **Croft-na-Caber** and **Portbane**

**Loch Treig**  See **Eadarloch**

**122  Loch Urr**
NX *c.* 755 843
NX78SE 6
Dumfries and Galloway Region - Stewartry and Nithsdale Districts
Kirkcudbrightshire - Balmaclellan ph. and Dumfriesshire - Dunscore ph.

Corrie mentions the discovery of 'Several canoes, in a poor state of preservation...near the Kirkcudbright outlet at Loch Urr'. The discovery was made 'many years' before 1927, and was apparently not recorded in detail.
  This extensive loch is situated in the Galloway Hills at an altitude of about 190m OD and in it there is an island-dwelling (NMRS NX78SE 2) which is probably of medieval date.

Corrie 1928, 292.

**Loch Venachar**  See **Portnellan Island**

**Loch Vennachar**  See **Portnellan Island**

**123  Lochar Moss**
NY *c.* 00 78
NY07NW 18
Dumfries and Galloway Region - Nithsdale District
Dumfriesshire - Tinwald ph.

A 'few years' before 1791 a logboat was discovered in a raised moss in Tinwald parish at an altitude of about 15m OD. It was discovered at a depth of between 4' (1.2m) and 5' (1.5m) during peat-digging and was noted as being 'of considerable size'. It was destroyed without detailed examination. See also nos. 79-80.

*Stat. Acct.*, i (1791), 160; Wilson 1851, 30-1.

**Locharbriggs**  See **Catherinefield**

### 124-8  Lochlea 1-5
Strathclyde Region - Kyle and Carrick District
Ayrshire - Tarbolton ph.

Within the area of the former loch of Lochlea or Lochlee there have been found five logboats, two oars, a paddle and a crannog. The loch was situated in the rolling clayland of the Cessnock valley at an altitude of about 120m OD and was drained about 1840. See also nos. A50-3.

1-2. NS c. 457 302   NS43SE 5

The initial drainage of the loch revealed two logboats, each of them about 12' (3.7m) long, 'on the south-west side of the crannog'. Their fate was not recorded.

Munro 1880, 30, 67.

3. NS c. 4574 3035   NS43SE 10

About 1878, a third logboat was dug up during further drainage operations about 100 yards (91m) N of the crannog and at a depth of at least 5' (1.5m). On discovery, it measured 10' (3.1m) in length and 2'6" (0.8m) in beam internally, and was 1'9" (0.5m) 'deep'. The sides were thinner than the flat bottom, which measured 4" (102mm) in thickness and was pierced by what were identified as nine thickness-gauge holes evenly disposed in two rows of four, with one at the bow; each measured 1" (25mm) in diameter and was found 'tightly plugged' with an unspecified material.

This logboat (fig. 25) is displayed in the Dick Institute, Kilmarnock; it was originally accessed as LC 244 but has been re-numbered as KIMMG: AR/D34. The underside was inaccessible at the date of visit. It has suffered slightly from splitting (one side having become detached) and more severely from warping, the bottom being warped upwards along the centreline and the sides inwards. The timber is pierced by four large knot-holes and numerous smaller examples. Practically the entire length of the boat survives but there is no certain differentiation between the two ends; one of them is, however, both broader and more rounded in form, so this is tentatively identified as the stern.

The boat measures 3.18m in length over all and up to 0.65m in beam; the assumed port side has probably measured about 0.15m in height but has become detached and is retained by nails in an inappropriate position. The probable starboard side survives to a height of 0.26m near the stern and 0.15m forward, but is lower amidships. The floor and sides measure about 60mm and 30mm in thickness respectively.

Eight probable thickness-gauge holes are arranged in equally-spaced pairs near the outer edges of the bottom of the boat; the even spacing of these holes may suggest an alternative explanation as location points for fitted ribs. They are oval after shrinkage and each measures about 30mm by 25mm; one of them retains its plug. A hole of unknown function and origin pierces the upper part of the surviving starboard side 0.9m from the stern; it measures 20mm in diameter externally and is angled slightly downwards into the interior where it is of slightly smaller diameter. There are the marks of an unidentified tool in the bottom of the boat on the starboard side amidships and in the port quarter.

On the basis of the current measurements and assuming the sides to be incomplete, the slenderness coefficient is 4.9. The remains of the boat are highly irregular in form and do not fall neatly into the morphological forms defined by McGrail. The probable stern appears to have been irregularly sub-rectangular in form, while the probable bow was apparently formed as a rounded point and the sides were vertical throughout. On the basis of this interpretation, the McGrail morphology code is 531:3x1:321 and the form dissimilar-ended. *December 1987*.

Munro 1880, 31, 67, 87.

4. NS 4575 3027   NS43SE 5

In 1878-9 Munro conducted excavations on the crannog. Among the objects found 'chiefly in the refuse-heap, and in the portion of debris corresponding to the area of the log pavement' he notes a 'Piece of wood like the back of a seat in a canoe'. This possible transom measures 2'4" (0.7m) in length by 9" (230mm) in breadth.

Munro 1880, 66.

5. NS 4575 3027   NS43SE 5

During the same excavations, Munro dug a central shaft to investigate the sub-structure of the crannog. Within the log pavement there was found 'part of a small canoe hollowed out of an oak trunk...evidently part of an old worn-out canoe...economised and used instead of a prepared log'. The surviving portion was 5' (1.5m) long and 1' (0.3m) deep; it varied in breadth between 1'2" (0.4m) broad at the end and 1'7" (0.5m) at the break. It is not recorded whether this boat was removed. The form of the boat cannot now be assessed and the incomplete nature of the remains precludes quantitative analysis.

Munro 1880, 49.

**Lochlee**  See **Lochlea**

### 129  Lochlundie Moss
NK c. 04 33
NK03SW 15

Grampian Region - Gordon District
Aberdeenshire - Slains ph.

Dalgarno notes the discovery of a logboat and 'broken oars' during peat-digging in Lochlundie Moss, near Ferny Brae and 'at a depth of several feet'. Lochlundie Moss is a raised peat-bog at an altitude of about 70m OD and the discovery was probably made before 1867, at which date the *Name Book* notes that the moss was 'pretty well drained'.

Dalgarno 1876, 1; Name Book, Aberdeenshire, No. 80, p. 11.

130-1 **Lochmaben, Castle Loch 1-2**
Dumfries and Galloway Region - Annandale and Eskdale District
Dumfriesshire - Lochmaben ph.

The discovery of two logboats is recorded in Castle Loch which is situated on the SE side of Lochmaben burgh and is dominated by its royal castle. The area is one of cultivated clayland and rounded relief at an altitude of about 45m OD. The discoveries of a possible crannog or 'artificial wooden structure', the 'remains of lake dwellings' and 'oaken mortised beams' (NMRS datasets NY08SE 9 and 30 respectively) have also been noted in the loch.

1. NY c. 090 811   NY08SE 50

The 'Ancient Boat or Canoe' that was exhibited to the Dumfriesshire and Galloway Natural History and Antiquarian Society in 1909 was probably a logboat. It was found near the outlet of the loch in unrecorded circumstances but was neither preserved nor recorded in detail. It measured 12' (3.7m) in length and 3' (0.9m) in beam, and was 'flat bottomed'. On the basis of these measurements, the slenderness coefficient was 4.

Contemporary newspaper accounts state that the 'ancient canoe' was found 'in a good state of preservation...in the outlet of the Castle Loch'. It was of 'black oak, rudely shaped and hollowed out'. The stern was 'shaped from the trunk in the same way as the bow' while 'artificial holes' were found in the bottom; 'wooden plugs or pins' (presumably thickness-gauges) were found in them but 'crumbled when removed'. The remains were 'split half way up the middle'.

A letter (dated 24 September 1909) from Mr. D. Fenton, then Town Clerk of Lochmaben, suggests an alternative location for the discovery 'about a mile below the Castle Loch at the outlet to the Annan river' (at NY c. 102 790). There is no supporting evidence for this suggestion.

The vessel was initially taken to Afton Lodge, Lockerbie, but its fate was not recorded. Although one of the logboats in store at Dumfries Museum has been identified as this discovery, these remains are more likely to be those of the example from White Loch (no. 154).

*Glasgow Herald*, 14 September 1909 and 25 September 1909; *Dumfries and Galloway Courier and Herald*, 18 September 1909; *TDGNHAS*, 2nd series, xxii (1909-10), 234; NMRS MS/678/118/1-10.

2. NY c. 08 81   NY08SE 51

The logboat that was revealed in 1949 when the level of the loch was lowered was found to measure 14' (4.3m) in length and 2'10½" (0.9m) in beam; evidence for a 'plank superstructure' (presumably in the form of washstrakes) was noted.

On the basis of the published account of the discovery, this is probably to be identified with one of the unlabelled logboats that are in store at Dumfries Museum, where the available records mention only the accession of one of the Castle Loch examples as DUMFM 1936.5(b) without specifying a particular discovery.

This boat (fig. 26) has been reduced by drying and splitting to a nearly-flat plank which measures 4.28m in length by up to 0.73m transversely and has been reinforced for display with iron bands. The underside was inaccessible at the date of visit. The bottom of the boat varies between 25mm and 40mm in thickness, being thickest near the edges which have become rounded through wear, and the sides are noticeably rounded in plan. The boat was possibly formed as a canoe with a slenderness coefficient of 4.9.

No evidence can be discerned for the proposed extension, but an unlabelled block of wood (measuring 1.5m by 130mm, and 80mm in thickness) which is stored with the boat may have formed part of the vessel or be associated with it in some way. *September 1987.*

*Archaeological News Letter*, 2 (1949-59), 68.

132-3 **Lochmaben, Kirk Loch 1-2**
NY *c*. 078 822
NY08SE 49
Dumfries and Galloway Region - Annandale and Eskdale District
Dumfriesshire - Lochmaben ph.

The discovery of two logboats is recorded in Kirk Loch which is situated on the SW side of Lochmaben burgh and in an identical topography to that of Castle Loch (nos. 130-1). These are apparently the two unlabelled logboats that are on wall-mounted display in Dumfries Museum and are accessed as DUMFM 1936.5(a) and (c), although there is no documentation to confirm the identifications which are deduced from the published account of the discovery. The undersides were inaccessible at the date of visit and the screw-holes noted in each boat probably result from their display mounting.

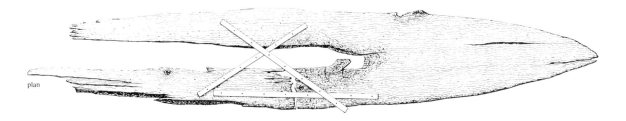

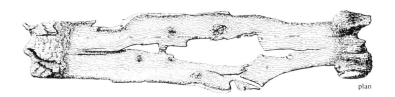

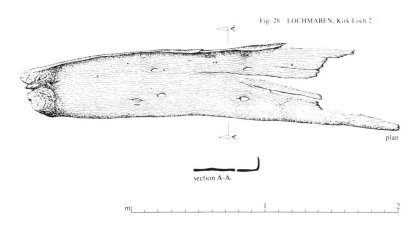

*Fig. 26. Lochmaben, Castle Loch 2 (no. 131). Logboat. plan of surviving remains, as currently retained. Scale 1:25.*

*Fig. 27. Lochmaben, Kirk Loch 1 (no. 132). Logboat. Plan of surviving remains. Scale 1:25.*

*Fig. 28. Lochmaben, Kirk Loch 2 (no. 133). Logboat. Plan of surviving remains. Scale 1:25.*

1. In December 1910 a logboat was revealed at a depth of 3' (0.9m) during the construction of a curling rink. It measured 8'5½" (2.6) in length and 2' (0.6m) in 'breadth', and was 'rounded on the bottom both inside and outside'. The timber was identified as probably being oak.

As displayed, the boat (fig. 27) measures 2.61m in length by up to 0.59 transversely. It has been worked from extensively-knotted timber without the use of thickness-gauges and has suffered grievously from warping, splitting and rotting so that it has been reduced to an extensively-perforated plank which is about 30mm thick, rises towards curved sides in places and has square-cut solid ends about 200mm high. On the basis of these measurements, the slenderness coefficient is 4.2. The McGrail morphology code is probably 112:1x2:112 and the form possibly that of a canoe variant. *December 1987.*

2. In March 1911 a second boat was revealed about 12 yards (11m) from (1). It measured 12'10" (3.9m) in length and 2'2" (0.6m) in beam, and the sides had been reduced to a height of about 8" (200mm). A 'small round hole' was noted 'at the broad end' and the timber was identified as 'oak'.

Parts of the boat disintegrated when it was raised and the surviving remains (fig. 28) comprise a flat plank measuring 2.84m in length by 0.65m transversely and between 10mm and 15mm in thickness, from which part of one side projects to a maximum height of 95mm. The 'broad end' has been lost. Warping and splitting have also affected the boat which was worked from knotted timber without the use of thickness-gauges. The surviving end has apparently been cut straight across and is roughly rounded in longitudinal section. The sides apparently taper inwards towards the end.

On the basis of the measurements recorded at the time of discovery, the slenderness coefficient was 5.9. The McGrail morphology code is 121:1x1:xxx or xxx:1x1:121 and the form cannot be determined. *December 1987.*
*TDGNHAS*, 2nd series, xxiii (1910-11), 321-2.


**Lochnameal**  See **Loch nam Miol**


**Lochside**  See **Buston**


134 **Lochspouts**
NS *c.* 258 058
NS20NE 25
Strathclyde Region - Kyle and Carrick District
Ayrshire - Kirkoswald ph.

About 1875 a logboat, which does not appear to survive, was discovered by Macfadzean during drainage operations on Lochspouts farm. The finder noted that it 'appeared small' and was 'roughly comparable' with the 'ruder forms' of the then-published examples from the River Clyde.

The exact location of the discovery and its relationship to the better-known crannog of the same name were not recorded. The area is one of undulating clayland and small lochs at an altitude of about 160m OD.

Munro 1882a, 2.


**Lochwinnoch**  See **Castle Semple Loch**


**Lotus Loch**  See **Loch Arthur**


135 **Mabie**
NX *c.* 95 71
NX97SE 17
Dumfries and Galloway Region - Nithsdale District
Kirkcudbrightshire - Troqueer ph.

In 1879 members of the Dumfriesshire and Galloway Natural History and Antiquarian Society inspected a 'fine old canoe' which had earlier been found during the drainage of Mabie Moss, an extensive area of peat-moss which was situated on the W edge of the valley of the lower Nith at an altitude of about 10m OD. This is presumably the 'oak boat' that Coles notes as having been found with 'other lacustrine relics' near the enclosure at Pict's Knowe (NX 953 721).

The boat measured about 14' (4.3m) in length by 3' (0.9m) transversely and its timber was identified as 'oak'. Plans were made for its preservation but do not appear to have been put into effect. Assuming the full length of the boat to have survived to that date, the slenderness coefficient was 4.7.

*TDGNHAS*, 2nd series, ii (1878-80), 28-9; Coles 1893, 123.


**Mabie Moss**  See **Mabie**


**Merton-mere**  See **Morton**


136 **Milton Island, logboat and paddle (possible)**
NS *c.* 41 73
NS47SW 13
Strathclyde Region - Dumbarton District
Dunbartonshire - Old Kilpatrick ph.

In the summer of 1868 a logboat was discovered in the River Clyde 'a little below Milton Island, near Dunglass'. It measured 22' (6.7m) in length and about 2'10" (0.9m) in 'breadth'; on the basis of these measurements the slenderness coefficient was 7.8.

The interior was 'well scooped out' and in it there were found six stone axes, a 'considerable piece of deer's horn' and an 'oaken war-club'. This latter object was possibly a paddle.

This discovery is probably to be identified with that noted by Currie 'opposite Dumbuck'. He identifies the timber as 'oak' and notes that the 'depth' could not be ascertained on account of the poor preservation of the sides. It was apparently recovered from the river but no longer survives.

Buchanan 1883, 77; *The Geological Magazine*, vi (1869), 37.


137  **Milton Loch**
NX 8388 7188
NX87SW  4
Dumfries and Galloway Region - Stewartry District
Kirkcudbrightshire - Urr ph.

In 1953 a crannog (Milton Loch 1) was excavated near the NW side of Milton Loch which is situated in the Galloway Hills at an altitude of about 125m OD. There was a possible dock on the SE side of the structure.

Excavation of this crannog revealed extensive timber framing supporting a single house with a central hearth, an access-causeway and a possible dock. An enamelled dress-fastener of about the 2nd century AD was found in excavation.

The following radiocarbon determinations have been obtained:

Structural pile of *Quercus* from
unstated location.
490 ± 100 bc  K-2027
Calibration yields multiple results
in the range 755-537 cal BC.

*Quercus* ard-stilt buried
beneath foundations.
400 ± 100 bc  K-1394
Calibration yields a date of 400 cal BC.

Pile from unstated location.
130 ± 50 bc  GU-2648
Calibration yields a date of 105 cal BC.

Although it has been suggested that the deposition of the dress-fastener post-dated the occupation, the possibility of a protracted or repeated occupation cannot be excluded.

What were possibly the remains of a logboat were found re-used as the threshold of the house, but were apparently neither recorded in detail nor preserved. This artifact measured about 4m in length and had been smoothed on one side but had suffered from rot on the other. Squared mortices had been cut into it at points 5'6" (1.68m) apart and the published plan indicates further working.

Piggott 1953, 137; Guido 1974.


**Monifieth**  See **Barry Links**


138  **Monkshill**
NJ *c.* 80 40
NJ84SW  16
Grampian Region - Banff and Buchan District
Aberdeenshire - Fyvie ph.

The Ordnance Survey note the discovery in June 1889 of what was probably a logboat at a depth of 7' in a peat-moss at Monkshill. It was kept in Fyvie parish church hall from 1908 until at least 1939, but is now lost. There are several remaining peat-mosses around Monkshill on the upper fringe of cultivated ground at an altitude of about 105m OD.

In 1978 there was recorded a local tradition that the 'Ancient British' boat was 'something like a coble', which may suggest that there was a stempost. The boat was 'not wholly complete', but the portion recovered measured about 8' (2.5m) or 10' (3.1m) in length.

Card index records of the former OS Archaeology Division (now in NMRS).


139  **Morton**
NX *c.* 89 99
NX89NE  16
Dumfries and Galloway Region - Nithsdale District
Dumfriesshire - Morton ph.

The *New Statistical Account* of the parish of Morton notes the discovery in the early eighteenth century of 'a boat cut out of one solid piece of wood, in the form of an Indian canoe' at 'the bottom of a moss not far from the old castle'. This is probably the discovery from 'Merton-mere' that is noted in the *New Statistical Account* of the parish of Kelton.

Morton Castle (NX 8908 9920) is situated at an altitude of 180m OD on the edge of upper Nithsdale and is bordered on the N, E and W by the loch of the same name.

Dumfries Museum records mention the transfer from the Grierson Collection of 'Fragments of Morton Loch dugout Canoe' under accession number DUMFM 1965.1814 but they cannot now be identified in the collections. These presumably formed part of the same discovery.

*NSA*, iv (Dumfries), 96; *NSA*, iv (Kirkcudbright), 155; Wilson 1851, 31; Ramage 1876, 312; Munro 1882b, 245; *TDGNHAS*, 3rd series, xi (1923-4), 111.

**Moss of Barnkirk**  See **Barnkirk**

**Moss of Knaven**  See **Knaven**

**Moss of Monkshill**  See **Monkshill**

**Newburgh-on-Tay**  See **Lindores**

**Oldcruivie Bank**  See **Lindores**

140 '**Orkney**'
Orkney Islands Area
Orkney - parish unknown

The remains of a logboat of impressive dimensions which is said to have been found in Orkney were on display in the Hunterian Museum, University of Glasgow (under accession number HM A.34) at the date of examination, but have since been removed to the Tankerness House Museum, Kirkwall, Orkney, where they remain in store under accession number THM 1994.21. There is no record of the date and circumstances of the discovery, but it was accessed into the Hunterian Collection before about 1900.

In view of the unusually large size of the vessel, the lack of detailed record of the provenance of the vessel, and the lack of oak trees of this size in the islands, an attribution to Orkney cannot be considered indubitable.

The boat (fig. 29) is now divided transversely into two unequal portions, having probably been divided by sawing before the onset of shrinkage. Although there is little difference in beam between the two sides of the cut, the profile and the thickness of the floor are so different as to suggest that the boat was formerly longer than the 5.3m that survives; an original length of about 5.5m may be postulated.

The timber has not suffered from warping but has many small knots and is greatly split. The sides do not survive to their full height and the remains are held together by modern bars, nails and bolts. Part of the bow has been replaced after breaking away, and two pieces of wood have been glued into the boat for an unknown purpose, and probably at a recent date.

The form of the boat is that of a lengthy triangle, with the greatest beam of 0.85m being found near the stern. The sides have apparently risen nearly vertically from rounded angles with the floor, which measures about 70mm in thickness. Thirty-six certain or probable thickness-gauge holes (spaced irregularly and at varying angles) pierce the floor and the sides. They vary in diameter between 10 and 40mm, and fifteen of them retain their wooden plugs.

The flared and rounded bow has been severely distorted by splitting and by cutting with a saw. The stern has been formed by a transom set in a groove which measures 40mm in breadth and 30mm in depth and displays distinct sharp-edged cuts which have probably been made with a metal chisel or similar tool. Displaced slightly to starboard on the forward side of the groove, there is a cut-out measuring about 50mm square which has probably held a supporting post. About 0.4m forward of the groove there is a rib left in the solid; this measures 80mm in breadth and is 30mm high on the forward side and about 10mm towards the stern.

Also awaiting removal to Tankerness House at the date of inspection, there were a further five pieces of timber (also fig. 29) which have apparently formed the remaining parts of the severely-damaged stern of this boat:

1. The transom of the logboat measures 1.04m across the top and is rounded in a roughly semi-circular form below; the greatest depth is 0.39m. It measures 50mm and 20mm in thickness across the flat top and the squared-off bottom respectively. The lower half of one side is bevelled.

2. A length of the starboard quarter including sections of the transom-groove and the false rib. It measures about 1.85m in length and up to 0.26m transversely and up to 170mm thick, but is partially protected by (museum-fitted) wooden side-boards so that the full length could be neither examined nor drawn; an iron peg protrudes at one point. There are two thickness-gauge holes; one of them pierces the main run of the side and measures 35mm by 30mm and 30mm in depth, while that near the end is inclined forwards and measures 40mm by 30mm and 80mm in depth.

3. A badly-split length of side (not drawn) which is labelled HM A34 and measures 0.74m by 0.24m and up to about 90mm in thickness. It is pierced by a thickness-gauge hole measuring 30mm by 20mm and 40mm in depth.

4. A badly-split length of curved side (not drawn) measuring 1.19m by 0.2m and about 40mm in thickness. It is pierced by a thickness-gauge hole measuring 25mm by 20mm and 55mm in depth.

5. A lump of timber (not drawn) of indeterminate form, measuring roughly 0.4m by 0.27m and about 190mm in thickness. It incorporates what is probably a length of the transom-groove.

On the assumptions that the boat originally measured 5.5m in length and that the height of the transom represents the original height of the side, the slenderness coefficient was

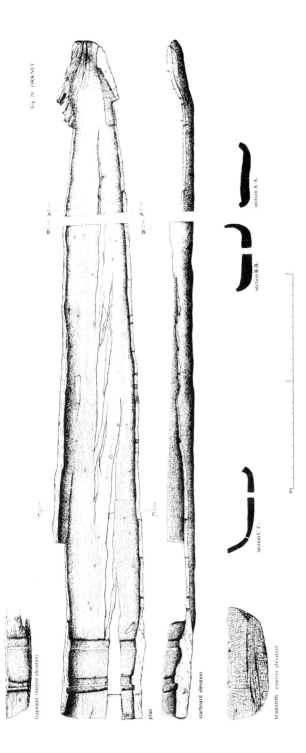

Fig. 29. 'Orkney' (no. 140). Logboat. Plan and starboard elevation of largest (divided) portion with interior elevation of detached fragment of stern, exterior elevation of transom, and sections on either side of division in the main portion and at about 70% of the (surviving) length from the bow. Scale 1:25.

6.5, the beam/draught coefficient was 2.2, and the displacement under standard conditions was 1.9 cubic metres. The McGrail morphology code is 44a4:2x4:233 and the form is dissimilar-ended. *August 1987 (main section) and February 1988 (fragments)*.

*The Orcadian*, 4 February 1988; NMRS dataset HY41SW 132.

### 141 Parkfergus
NR c. 66 21
NR62SE 3
Strathclyde Region - Argyll and Bute District
Argyll - Campbeltown ph.

In 1790 the drainage of the former Loch Sanish revealed what was probably a logboat; it does not appear to survive. The records of the former Ordnance Survey Archaeology Branch locate the discovery at NR c. 6650 2130, but the evidence for this is unstated.

The location is in South Kintyre, in a flat area at an altitude of about 5m OD on the S side of Machrihanish airfield, and is probably based on underlying marine silt.

M'Intosh 1861, 15; McInnes 1935, 22, no. 28.

### 142-3 Port Laing 1-2
NT c. 133 811
NT18SW 25
Fife Region - Dunfermline District
Fife - Inverkeithing ph.

About 1857 two 'large canoes' were found 'imbedded in the sand' during whinstone quarrying at the S end of Port Laing. They were broken up and the 'small specimens' remaining were donated to the Society of Antiquaries of Scotland; these cannot now be identified in the collections of the Royal Museum of Scotland.

Chalmers 1844-59, ii, 388-9.

### 144 Portbane
NN c. 768 448
NN74SE 14
Tayside Region - Perth and Kinross District
Perthshire - Kenmore ph.

About 1977 sports divers found what was probably a logboat in about 8m depth of water in front of the hamlet of Portbane near the NE end of Loch Tay. It was neither recorded nor recovered. See also no. 24.

*Pers. comm.* Mr C Aston.

### 145 Portnellan Island
NN 5925 0615
NN50NE 2
Central Region -Stirling District
Perthshire - Callander ph.

In 1913 what was possibly the bottom of a logboat was found whilst what was probably a crannog was under examination during conditions of low water level. The form of the timber was not recorded but it measured 14' (4.3m) in length by 2' (0.6m) transversely and was found with other 'waterlogged planks' embedded 'partly in the loose stones on the west...side of the island'. There is no record of its removal, and it may remain *in situ*.

The crannog is situated near the E end of Loch Venachar or Vennachar, which occupies one of the glaciated valleys of the Trossachs at an altitude of about 80m OD.

Fleming 1915, 341-2.


### 146-7 Redkirk Point 1-2
NY *c.* 302 650
NY36NW 14
Dumfries and Galloway Region - Annandale and Eskdale District
Dumfriesshire - Gretna ph.

In 1954 and 1956 respectively two possible logboats were exposed by marine erosion in the estuarine clay of the N shore of the Solway Firth. Neither of them was preserved or recorded in detail, and they may have been mis-identified trees.

1. The first discovery contained fragments of rusted iron, one of which was possibly a sword.

2. The second was found 'a few feet' away.

*DES*, (1954), 8; *DES*, (1956), 14.


### 148 River Carron
NS88SE or NS98SW
Central Region - Falkirk District
Stirlingshire - Grangemouth, Falkirk or Larbert ph.

In May 1726 erosion revealed a logboat in the 'washings of the River Caron' and at a depth of about 15' (4.6m) beneath several layers of what were presumably natural deposits. The location of the discovery was not recorded but it was probably in the extensive estuarine carseland along the lower course of the river.

Sir John Clerk of Penicuik noted that the boat measured 36' (11m) in length, 4'6" (1.4m) in beam and 4'4" (1.3m) in depth; the sides were 4" (102mm) thick. The stem was 'sharp' and the stern 'square' while the timber was knot-free 'oak' which had been 'finely polished'. On the basis of these measurements the slenderness coefficient was 7.9, the beam/draught coefficient was about 1 and the displacement under standard conditions was about 11.8 cubic metres. These figures indicate a large logboat of narrow and dissimilar-ended form (presumably with a solid stern) and having a considerable speed potential.

Although the account of the discovery does not describe the stratigraphy in detail, it is likely that the boat was found in the carse clay that was laid down during the marine transgression between 6000 and 4000 bc (Morrison 1980, 102, 104); it thus probably dates from the late Boreal or early-to-mid Atlantic periods.

This discovery is noted by Munro as being found in the Carse of Falkirk, but is distinguished from that at Falkirk (no. 45).

*Bibliotheca Topographica Britannica*, (1790), ii, 241-2; Munro 1898, 272; NMRS dataset NS88SE 88.


### 149 'River Clyde'

On display in the Hunterian Museum (under accession number HM A.33) there is an impressive logboat for which there are no records of provenance or discovery. This boat may be equated with any of those that are said to have been preserved but cannot now be located, including particularly Buston 3 (no. 13).

The boat (fig. 30) has been worked from a whole log of timber with few but large knots and has suffered from splitting (most noticeably around the bow) and warping which has displaced the bottom of the boat downwards along the sides. The vessel measures 6.35m in length over all by up to 0.94 transversely and the port side survives to a height of about 550mm above the bottom.

The bow is of rounded point form and is rounded both internally and externally in longitudinal section, but is flat with rounded corners when viewed transversely. The stern has been formed by a transom set at an angle to the centreline in a groove which measures 35mm across and 20mm in depth, and survives all the way up the port side; the starboard side is missing at this point. The transom is set in a raised section, and immediately forward of its groove there is a single irregular mark which was possibly made during recovery operations.

The sides of the long midships section are nearly parallel and rise vertically from the bottom of the boat, except that the port side is rounded externally amidships. Both sides and bottom measure about 50mm in thickness.

Along the upper surfaces of the two sides there are a total of fourteen dowel-holes which measure up to 30mm in diameter and 60mm in depth. One of them retains a projecting dowel about 20mm high and another has its dowel broken off level with the upper surface of the sides. On the starboard side there are eight holes roughly spaced at intervals of between 0.6m and 0.8m, but the six holes on the port side are less evenly spaced. Three of them are set within 550mm near the bow, but there are none at all near the stern. Although these features might be seen as evidence of washstrakes or extended construction, their setting into an uneven upper surface suggests that they had a structural function, being intended to retain sections of the side which had broken or split away.

At a point some 2.7m from the bow, the exterior of the port side has been partially cut away to about half the thickness of the side over a length and depth of about 80mm for an unknown purpose and possibly during recovery operations. About 1.65m from the stern there is a major declivity in the same side which apparently owes more to the presence of a massive knot with its associated splitting than to any constructional feature.

In the floor of the boat there are four thickness-gauge holes which measure between 20mm and 30mm in diameter and are spaced irregularly along the centreline. The two nearer the bow retain their plugs while the two nearer the stern may be unfinished or have retained the plugs set deep in the holes.

Set transversely across the boat at about 2.6m and 4.8m from the bows, there are two false ribs which continue up the inner sides; each of them measures about 90mm in breadth and about 15mm in height.

On the assumption that the port side survives to its full height, the slenderness coefficient is 6.4, the beam/draught coefficient is 1.7 and the displacement is about 1.96 cubic metres under standard conditions. Assuming the longitudinal profile of the surviving side to accurately represent that originally constructed, the McGrail morphology code of the boat is 44c1:111:322 and the form is dissimilar-ended. *October 1987.*

*Pers. comm.* Mr J Hunter; NMRS dataset NS47SE 64.

**River Conon** See **Dingwall**

**River Spey** See **Garmouth**

150 **River Forth**
NS *c.* 88 91
NS89SE 78
Central Region - Clackmannan or Falkirk District
Clackmannanshire or Stirlingshire - Alloa or Airth parish

In about 1836 or 1839 a 'canoe formed out of a single oak log' was found 'adrift in the Forth below Alloa', where it was 'supposed to have been lost by a Norwegian vessel'. It was brought upriver to Causewayhead where it was 'used for short trips on the river by the juvenile Causewayheadians' until it 'went missing'. It is said to have sunk in the area of the Cambuskenneth logboat (no. 14), with which it was conflated in a later newspaper account.

No detailed record was apparently made and many of the details cited appear dubious at best, but the essential

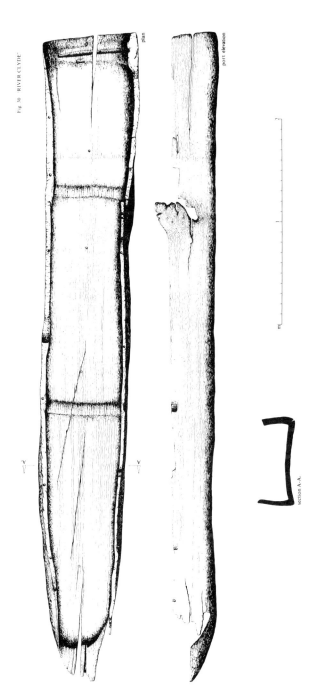

*Fig. 30. 'River Clyde' (no. 149). Logboat. Plan, port elevation and section at 34% of length from the bow. Scale 1:25.*

particulars appear to be consistent with the discovery of a logboat.

*Stirling Observer*, 21 May 1874.


151 **'River Tay'**
Tayside Region - Perth and Kinross District
Perthshire - Perth, Kinnoull or Kinfauns ph.

In Roxburgh District Museum, Hawick (under accession number HAKMG: 4101) there is a fragment of wood which is labelled as a 'Fragment of prehistoric canoe found in River Tay near Perth'. This split, facetted and possibly axe-marked peg of fine-grained wood measures up to 160mm by 40mm by 30mm; it has evidently been sawn across at the ends and one face is apparently bevelled.

This fragment cannot be classified by function or origin, or identified as any part of a logboat. It is noted as a separate discovery but may conceivably have formed part of any of the boats from the River Tay, namely Errol (nos. 37-8), Friarton (no. 50), Lindores (nos. 87-8), Sleepless Inch (no. 152) or Perth, Saint John Street (no. A58). *May 1994.*

*Pers. comm.* Mrs R. Capper; NMRS dataset NO12SW 210.


**St Columba's Loch** See **Loch Chaluim Chille**


**Sanquhar Loch** See **Black Loch**


152 **Sleepless Inch**
NO *c.* 146 220
NO12SW 211
Tayside Region - Perth and Kinross District
Perthshire - Rhynd or Kinfauns ph.

In June 1848 a logboat of 'oak' was 'taken out of the bed of the Tay, at Sleepless Island'. One end was missing but the remains measured about 22' (6.7m) in length. It was intended to preserve the boat, but this does not appear to have been achieved. See also no. 151.

*Perthshire Courier*, 22 June 1848.


153 **Stirling, King Street**
NS *c.* 795 934
NS79SE 168
Central Region - Stirling District
Stirlingshire - Stirling ph.

In 1865 a 'portion of a canoe' was found (probably during building operations) at the premises of Messrs. Menzies, drapers, and at a depth of about 3'6" (1.1m). It was said to be 'of black oak and in one piece'. An iron ball was found in it.

King Street extends from NS 7958 9343 to NS 7967 9334 and the location is at an altitude of between 19m and 25m OD on the lower slope of a crag-and-tail formation.

Morris 1892, 23-4.


154 **White Loch**
NX *c.* 103 608
NX16SW 31
Dumfries and Galloway Region - Wigtown District
Wigtownshire - Inch ph.

A logboat was discovered about 1870 in White Loch, which has also been known as Loch of Inch and is situated in drumlin country at an altitude of about 20m OD. Neither the boat nor the circumstances of its discovery were recorded in detail, but it was found near the channel between the shore and the island which was the site of the 'Manor Place of Inch', a medieval house of the Earls of Cassillis.

In store at Dumfries Museum there are the unlabelled remains of a logboat which was identified by Mrs Grant as Lochmaben, Castle Loch 1 (no. 130) but is more probably the White Loch discovery. Museum records note the transfer from the Stair Collection of a logboat from 'Lochinch Castle' which bears the accession number DUMFM 1964.139, and of which no other remains are evident in the collection. The underside of the boat was inaccessible at the date of visit.

The remains (fig. 31) comprise one end and the midships section of a logboat worked from knotted timber. It has been reduced to a flat plank from which project the end and part of one side. The other end has broken away and been lost, so that the remains now measure only 3.94m in length. Iron bands have been fitted to retain the severely-split remains in position.

The surviving end (which was more probably the bow) is of rounded point form, survives to a height of up to 140mm and is rounded both externally and internally in longitudinal section. The midships section is parallel-sided and measures about 0.5m in beam. The sides were probably slightly flared and one of them survives in part to a height of 200mm. The bottom measures about 30mm in thickness along the centreline, and between 10mm and 20mm near the sides.

On the evidence of the surviving remains, the slenderness coefficient was in excess of 7.9. The McGrail morphology code is 131:1x3:xxx or (less probably) xxx:1x3:131 and the form could not be ascertained from the surviving evidence. *December 1987.*

Dalrymple 1872, 389.

## II  Gazetteer of Logboat Discoveries

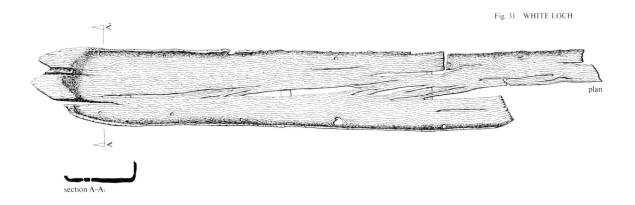

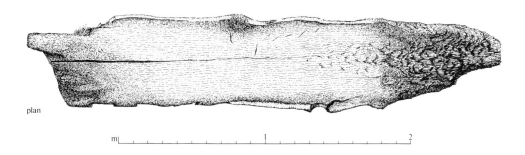

*Fig. 31. White Loch (no. 154). Logboat. Plan view of surviving remains and section near bow. Scale 1:25.*
*Fig. 32. Daviot timber (no. A15). Unworked timber, formerly proposed as possible logboat. Plan view, for comparison with accepted logboat remains. Scale 1:25.*

# III  Examples of Related Artefact-types

**An Cnatharan**  See **Plockton, keg**

**A1-4  Ardgour 1-4, bowls**
Unlocated
NM96SE 6
Highland Region - Lochaber District
Argyll - Ardgour ph.

Held in store at the Royal Museum of Scotland (under accession numbers NMS ME 65-8) there are probable remains of a group (possibly a hoard) of wooden objects which were found (apparently during peat-cutting) before 1871 in a 'moss' near Ardgour House. It is unclear from what type of timber they have been worked, but alder or birch appear the most probable.

1. NMS ME 65 is a turned bowl with a vertically-set loop handle, which is now broken. As tentatively reconstructed by Henshall, it has measured 6" (152mm) in diameter by 2½" (64mm) in depth. The original account of the discovery notes that it had been 'mended with two bronze clasps secured by small rivets' but only one clasp can now be identified. This is neatly-made, evidently of bronze and secured by a single rivet over the rim on the opposite side from the handle; its identification as a repair feature is confirmed.

The vessel is closely-paralleled by one from Armagh, Ireland and has been radiocarbon-dated to 155 ± 70 ad (OxA-2415); this determination may be calibrated to about 195 or 230 cal AD.

2. NMS ME 66 is a carved bowl of elliptical form with two opposed loop handles, one of which is apparently broken. It measures 10.1" (257mm) 'long' by about 7.5" (191mm) 'broad' and 4.5" (114mm) 'deep', has been 'crudely cut from the solid', and has been radiocarbon-dated to 60 ± 70 ad (OxA-2416), which determination may be calibrated to between about 81 and 102 cal AD.

3 and 4. NMS ME 67 and 68 are fragments of carved bowls with carved handles set horizontally. Only the handle and part of the rim survive in each case, while ME 68 is fragmentary.

This group is immediately significant as indicating the contemporary use of turned and carved vessels. The turned example are seen by Earwood as exemplifying a relatively common group with parallels at Dalvaird Moss (no. A14), Loch Laggan (no. A47) and widely in Ireland (her 'County Armagh' type). On the basis of the few available dates, this has been a long-lasting type, in use between at least the third century BC and the second century AD in Ireland and between the first and sixth centuries AD in Scotland. These wooden bowls appear to have had a similar distribution to their bronze counterparts, the lack of any recorded examples from Northern England being possibly a reflection of the lack of archaeological recording in that area. She sees the carved vessel of elongated form (ME 66) as indicating a date for the type in the first few centuries AD, and notes parallels in Ireland and at Talisker Moor (no. A68). *November 1994*

*PSAS*, ix (1870-2), 220-1; Maxwell 1951, 164; Earwood 1990b, 37-9, 44; Hedges, Housley, Bronk and van Klinken 1991, 128; Earwood 1993a, 62, 64-5, 67, 219, 265; Crone 1993b, 274.

**A5  Arisaig, Loch nan Eala, timbers**
NM 6680 8585
NM68NE 2
Highland Region - Lochaber District
Inverness-shire - Arisaig and Moidart ph.

About 1856 a crannog was revealed by the drainage of Loch nan Eala, a freshwater loch situated about 1km SE of Arisaig village. The subsequent excavation of the crannog revealed a timber 'raft' measuring about 50' (15.2m) square and 4' (1.2m) thick. which was retained around the sides by two rows of sloping timber 'stays'. (0.76m) in circumference.

During this excavation there were found 'Two great logs...nicely rounded off at the end', which were apparently allowed to disintegrate. Each had a dugout hollow measuring about 2 or 3" (51 or 76mm) in depth and 4" (102mm) in breadth.

In the absence of a comprehensive description of these objects, their nature and function remain unclear. The restricted dimensions of the dugout hollows indicate that they were probably not logboats, although the possibility that they were unfinished examples cannot be entirely ruled out. The 'boat' from this area that is mentioned in an account of the Acharacle logboat (no. 1) is most probably one of these logs.

Mapleton 1868, 518; Blundell 1911, 359-62; NMRS MS/47/1.

### A6 Bailemeonach, Mull, trough
NM 6577 4153
NM64SE 7
Strathclyde Region - Argyll and Bute District
Argyll - Torosay ph.

In 1936 a 'Trough-like object' from Fishwick Bay was donated to the National Museum of Antiquities of Scotland (now the Royal Museum of Scotland). It was allocated accession number MP 580, but cannot be identified in the collections. This object may, however, be one of the two unprovenanced timber artifacts that are stored in the Museum (nos. A21-2).

The object from Fishwick Bay is described as being 'dug out of the trunk of an oak tree, open at one end and closed at the other', and measured 4'3" (1.3m) in length, 1'7½" (0.5m) in height and 1'7" (0.48m) in external breadth. The bottom was pierced near the closed end by a hole which has cross-diameters of 1' (0.3m) and 11" (0.28m). On the basis of the measurements cited, the capacity appears to have been somewhat under about 98 litres.

In 1972, officers of the Ordnance Survey interviewed a witness to the discovery and established its location at an altitude of 20m OD and at the head of a small valley about 1km from the NE coast of Mull. Although its discovery near to a stream may suggest that it formed part of a watermill or similar installation, it is said to have been found at a depth of about 6' (1.8m) during drainage works; it was most probably a bog butter trough. Earwood has suggested that it is late prehistoric in date.

*PSAS*, lxxi (1936-7), 23; Earwood 1993a, 50, 275.

### A7 Bankhead, mill-paddle
NX 937 847
NX98SW 70
Dumfries and Galloway Region - Nithsdale District
Dumfriesshire - Kirkmahoe ph.

In store at the Royal Museum of Scotland (under accession number RMS PD 12) there is a paddle-blade which was found before 1956 during drainage operations at a point about 300m E of Bankhead farmsteading. The area is situated on the edge of the haughland of the middle Nith valley, and at an altitude of about 45m OD.

The blade is of oak, and measures 1' 3½" (0.39m) in length by up to 3.85" (90mm) in width and 1.05" (32mm) in thickness. The ventral surface is concave in both planes and has a slight twist; the distal end is rounded and the proximal end is tapered around a distinct collar.

This form of paddle has been compared with that commonly used in Irish horizontal mills and clearly demonstrates the differences between mill-paddles and those used for waterborne propulsion. *August 1987*

Maxwell 1956.

**Barhapple Loch, paddle** See **Barhapple Loch 2**

**Bowling, paddle (possible)** See **Bowling 1-2**

### A8 Bunloit, Glenurquhart, keg
NH *c.* 50 25
NH52NW 10
Highland Region - Inverness District
Inverness-shire - Urquhart and Glenmoriston ph.

Held in the collections of the Royal Museum of Scotland (under accession number RMS SHC 7) there is a mass of bog butter with fragments of the timber vessel that contained it. The circumstances of the discovery are not recorded but the accession dates from 1904. Bunloit farm is at grid reference NH 504 252, above the NW shore of Loch Ness and 4.2km SW of Urquhart Castle.

As currently held in store, the remains appear to be those of a lump of bog butter (measuring about 0.46m in length by 0.29m in diameter) around which the remains of a timber container are retained by wire binding (which is presumably of recent date). The timber (which was possibly alder) has split so severely that it presents the superficial impression of a stave-built container, but is shown by detailed examination to be of monoxylous (presumably carved) form. Both base and lid are missing and the ends of the sides have rotted so as to obliterate any trace of a binding-groove or similar feature. The top of the bog butter mass is domed and the capacity of the vessel appears to have been about 141 litres. *November 1994*

*PSAS*, xxxix (1904-5), 246-7; RMS Accession Register: typescript continuation catalogue.

**Buston, oar** See **Buston 1**

## A9  Cairngall, log-coffins
NK c. 04 47
NK04NW 3
Grampian Region - Banff and Buchan District
Aberdeenshire - Longside ph.

The *New Statistical Account* notes the discovery of 'two oak coffins or chests' when a 'tumulus of moss' (presumably a barrow) was removed on the 'estate of Cairngall'. Cairngall House is at NK 042 473.

One of the coffins was 'entire' but the other was incomplete; at least part of one of them was transferred to the Arbuthnott Museum, Peterhead, but had probably been lost by 1968. Each had been hollowed from a 'solid tree' and measured 'seven feet' (2.1m) 'by two feet' (0.6m). Their sides were parallel and their ends rounded, and each had two 'projecting knobs to facilitate their carriage'; the well-preserved bark was still to be seen. No bone or organic contents survived, but they lay E-W and were covered by 'slabs of wood'.

The discovery is noted by Wilson who attributes it to his 'iron period'. Childe, however, places it within his 'Food-vessel complex' on the basis of English comparanda, and the evidence of the coffin from Cartington, Northumberland (section IV.9.1, below) appears to support this assertion.

*NSA*, xii (Aberdeen), 354-5; Wilson 1851, 462; Anderson and Black 1888, 366; Abercromby 1905, 181; Childe 1946, 119.


## A10  Cairnside, mill-trough
NW 979 709
NW97SE 8
Dumfries and Galloway Region - Wigtown District
Wigtownshire - Kirkcolm ph.

In 1885 a 'curious dug-out trough like a canoe' was revealed during field-drainage operations 'among the remains of an ancient mill' which was apparently situated 390m NNW of Cairnside farmhouse and at an altitude of 25m OD on the bank of an un-named and un-navigable stream.

The trough was found beneath a layer of clay at a depth of about 2' (0.6m) and measured 9'5" (2.9m) in length, 1'11" (0.6m) and 1'5" (0.4m) in external and internal breadths respectively, and 1'3" (0.4m) in depth. The capacity was thus about 460 litres.

The lower end was cut square across and pierced by two holes which tapered from 5" (125mm) in diameter internally to 3" (76mm) externally. Each of these holes was covered by a small patch of wood which measured 1¼" (31mm) in thickness, was pierced by a smaller hole, and was retained in place by ten treenails of different sizes. Three of the treenails noted on each of the longer sides of each patch measured 1" (25mm) in thickness, and the other two measured half of that. The heads of the larger examples were 'bent at a right angle like a walking-stick' in a manner similar to that used to retain a rib of the Dernaglar Loch logboat (no. 27) and another (which was possibly derived from a logboat) found in the structure of the crannog at Barhapple Loch (nos. 4-5). The perforated end of the trough rested in a groove cut into the outer edge of a heavy squared beam which was adjacent to a floor of thick oak planks and close to two granite millstones, each of which measured 3' (0.9m) in diameter.

The circumstances of the discovery in the remains of a watermill and out of proximity to navigable water make it improbable that this object was a logboat and confirm its (unpublished) identification as a 'trough for conducting water to mill wheel'.

Wilson 1895, 72-3; RCAHMS 1985, 37, no. 235; MS. notes in RMS: SAS 457.


**Closeburn, paddle**  See **Kirkbog, paddle**


## A11  Cnoc Leathann, Durness, trough
NC 3871 6384
NC36SE 14
Highland Region - Sutherland District
Sutherland - Durness ph.

In 1969 a trough was found in moorland on the E side of the Kyle of Durness, and at an altitude of about 70m OD. After recovery (in a fragmentary and waterlogged condition) it was re-assembled and consolidated with acetone-rosin. It is stored at the Royal Museum of Scotland under accession number NMS SHC 9. It was found to contain an irregular pitted lump of bog butter (NMS SHC 10) which measures 340mm in length and has been dated by radiocarbon to $1110 \pm 80$ ad (OxA-3010), which determination may be calibrated to about 1212 cal AD.

The trough has been worked from a near-complete section of an oak log and now measures 730mm in length over two prominent flat lugs which are cut straight across the ends. The end walls measure up to about 36mm in thickness and the enclosed cavity is about 480mm long; the object measures up to about 285mm in breadth externally and 186mm internally. The exterior is rounded and the cavity is square-cut at the corners although the sides bow out slightly. The internal depth is about 111mm and the contained volume about six litres.

The vessel generally displays a high standard of workmanship, both interior and exterior surfaces being smooth. Among scratchmarks within the interior there are the marks of what has probably been a metal chisel about 20mm broad. The exterior surface was not accessible at the date of inspection. *November 1994*

RMS accession register: typescript continuation catalogue; Close-Brooks 1984, 580; Hedges, Housley, Bronk and van Klinken 1992, 144-5; Earwood 1993a, 12, 100, 101, 274.

**A12 Craigie Mains, oar**
NS *c.* 405 316
NS43SW 18.01
Strathclyde Region - Kyle and Carrick District
Ayrshire - Craigie ph.

An oar was found before 1895 during drainage operations to the SW of Craigie Castle, where a possible crannog was also noted. It was not recorded in detail and has not survived.

Smith 1895, 129; RCAHMS 1985b, 11, no. 43.

**Creag a' Chapuill** See **Rubh' an Dunain, oar or paddle (possible)**

**Culsalmond** See **Williamston, log-coffin**

**A13 Cunnister, Yell, Shetland, trough**
HU *c.* 52 96
HU59NW 14
Shetland Islands Area
Zetland - Yell ph.

In 1887 an 'oblong wooden vessel' was found at a depth of 3' (0.9m) in a peat moss at Cunnister or Gunnister which is situated on the E side of Basta Voe in the NE part of Yell. It is recorded as being in store at the Royal Museum of Scotland under accession number NMS SHC 6.

The vessel was described as being 'somewhat like a large nut in appearance' and 'covered by a thin covering of fibre, evidently the inner bark of some tree'. It measured 1'8" (0.5m) in length over all by 1' (0.3m) transversely, and 7" (178mm) in thickness. The internal cavity contained a 'mass' of bog butter.

Held in store at the Royal Museum of Scotland there is an (unlabelled) object which may be identified as this trough. After (presumably) shrinkage, it measures about 0.64m in length by 0.25m in breadth and 0.16m in depth and has been worked from a length of gnarled and knotted timber which is roughly-square in section and has been split in places; the shaping of the exterior has been minimal and the light weight of the piece appears to confirm the proposed timber identification. Internally, the piece is roughly squared at the ends and rounded in section to produce a cavity of roughly cylindrical form measuring about 0.48m in length by 0.19m in breadth and 0.11m in depth; the volume thus appears to have been about 1 litre. No lid appears to survive and there is no indication of one having existed. Toolmarks of indeterminate form are visible within the interior, and there appears to have been some rotting of the timber. *November 1994*

Macadam 1889, 433; Ritchie 1941, 13, 16; Earwood 1993a, 276; Crone 1993b, 274.

**Dalrigh** See **Oban, log-coffin**

**Dalswinton** See **Bankhead, mill-paddle**

**A14 Dalvaird Moss, Glenluce, bowl**
NX *c.* 40 72
NX47SW 10
Dumfries and Galloway Region - Wigtown District
Kirkcudbrightshire - Minnigaff ph.

Held in store at the Royal Museum of Scotland under accession number NMS SFA 6 (having formerly been numbered ME 70) there is a wooden bowl (or 'cup') which was found 'under a large wooden dish' (which does not appear to survive) in Dalvaird Moss. It has a vertically-set handle and is of similar form to one of those (no. A1) from Ardgour. It has been dated by radiocarbon to 265 ± 70 ad (OxA-2414), which determination may be calibrated to between about 339 and 372 cal AD.

*PSAS*, ix (1870-2), 357; Black 1894, 47; Maxwell 1951, 164-5; Earwood 1990b, 39, 44; Hedges, Housley, Bronk and van Klinken 1991, 128; Earwood 1993a, 64, 65, 67, 272; Crone 1993b, 274.

**A15 Daviot, timber**
NH 714 407
NH74SW 25
Highland Region - Inverness District
Inverness-shire - Daviot and Dunlichity ph.

In March 1977 an object which was tentatively identified as a logboat was discovered during road construction across afforested moorland about 1.4km WNW of House of Daviot, at an altitude of about 220m OD and at least 1.6km from the nearest possible navigable water. It is in store at Inverness Museum under accession number INVMG 983.165.

The object (fig. 32) bears a superficial resemblance to a logboat but its present form is apparently the result of rotting or similar natural processes. It is probably of no archaeological significance but the possibility that it is an unfinished or crudely-worked log coffin, trough or similar artifact cannot be entirely excluded. Examination by a tree scientist is reported to have indicated that the form of the object is typical of rotted oak trees and confirmed that no human activity need be indicated.

After shrinkage, it measures 3.24m in length by up to 0.62m transversely. It has been formed from a tree-trunk, the bark of which has not been removed, and from which project several branches. There are several large knots in the timber, which has suffered from both warping and splitting. The supposedly-hollowed central section measures 0.16m in maximum depth, beneath which there is up to 0.2m thickness of unworked timber. Neither thickness-gauge holes nor toolmarks are to be seen. Both the narrow 'bow' and the 'stern' are ragged and unformed; the former has been bent downwards by warping while the latter rises about 70mm above the interior and bears no evidence of a transom. *September 1987*

*Pers. comm.* Miss J. Harden.


**Dowalton Loch, paddle** See **Ravenstone Moss, paddles**


**A16 Drumcoltran, timber**
NX 873 682
NX86NE 21
Dumfries and Galloway Region - Stewartry District
Kirkcudbrightshire - Kirkgunzeon ph.

In 1984 a split oak log was discovered during drainage work about 400m E of Drumcoltran Tower, and initially entered the press as the remains of a possible logboat. Further examination revealed no evidence of human workmanship or of antiquity. The timber was, nonetheless, placed in a tank for conservation.

*Scotsman*, 10 May 1984; *pers. comm.* Mr P. Hill; NMRS A41544-53/po.


**Drumglow** See **Dumglow, log-coffin**


**A17 Dumglow, log-coffin**
NT 0759 9649
NT09NE 4
Tayside Region - Perth and Kinross District
Kinross-shire - Cleish ph.

The cairn that is situated within the fort on the summit of Dumglow (or Drumglow), and at an altitude of 379m OD, measures 50' (15.3m) in diameter and 5' (1.5m) in height. In 1904 it was excavated by trenching and found to be of earthen construction with a capping of stones.

In the centre (aligned E-W at a depth of 2m below the highest point) there were found the decayed remains of a 'hollowed-out tree-trunk of oak' which measured 7'1" (2.2m) in length by up to 11" (279mm) in breadth. Although neither human remains nor grave-goods were found, this was identified as a log-coffin.

Childe suggests an attribution to the Bronze Age on the analogy of the comparable English examples of Food Vessel association that are discussed below (section IV.9.1).

Abercromby 1905, 179-81; RCAHMS 1933, 290, no. 550; Childe 1946, 119; Feachem 1977, 75.


**Durness** See **Cnoc Leathann, Durness, trough**


**A18 Eadarloch, trough**
NN *c.* 347 768
NN37NW 3
Highland Region - Lochaber District
Inverness-shire - Kilmonivaig ph.

The smaller of the two timber objects from Eadarloch that are displayed in the West Highland Museum, Fort William, was probably a bog butter trough. The circumstances and topographical situation of its discovery are discussed under no. 36 above.

As currently displayed (fig. 43), the object measures 1.67m in length over all, up to 0.3m in breadth and 0.26m in depth externally. The upper surface is continued over the square ends in the form of horizontal projections. The hollowed portion is noticeably rectangular on plan and square in section, measuring 1.39m in length and 0.25m in breadth at the top of the sides. The flat bottom is 0.16m below the tops of the vertical sides and the ends slope inwards, making the capacity about 50 litres. The timber, which is prominently-grained, has previously been identified as oak and has suffered slight splitting; the sloping ends of the hollowed section display medullary rays and there are two knots in the bottom. The pronounced toolmarks in the lower part of one of the sloping ends probably result from undercutting strokes with an adze or similar metal tool, but the exterior has not been smoothed and there are no thickness-gauge holes. The McGrail morphology code is 151:111:151.

The function of this object has been much discussed. Ritchie suggested, on the basis of Irish parallels, that it was possibly a child's boat, a portable food-trough or a logboat used for the transport of building materials, while Sayce compares it with other recorded cooking-troughs. The presence of two probable portage-handles is unparalleled but need not be conclusive evidence for it not being a logboat. The small size and lack of clearance-gauge holes, however, make such an identification improbable. The narrow and elongated shape is inappropriate for a log-coffin but suitable for a cooking-trough. In the absence of convincing parallels of similar form, the possibility that it was a logboat of unusual form cannot be entirely ruled out. *July 1987*

Ritchie 1942, 57-8, 77-8; Sayce 1945; Earwood 1993a, 102, 274.

A19  **Easter Oakenhead, boat**
NJ *c.* 24 68
NJ26NW  7
Grampian Region - Moray District
Morayshire - Drainie ph.

About 1833 an 'ancient boat', which was probably not a logboat, was found during the ploughing of land reclaimed from Loch Spynie. It measured 30' (9.1m) in length and the stern was 'quite round'. Within it there were found 'ribs' of 'oak'.

Loch Spynie was an arm of the Moray Firth until it was cut off by the growth of shingle-bars in medieval times. The boat probably dates from before the late 18th century, when the Spynie Canal was dug to drain the area and reduce the loch to its present size.

*Stat. Acct.*, x (1794), 623-4; *The Gentleman's Magazine*, (1834), i, 95, reprinted in Gomme (ed.) 1886, i, 55.

A20  **Edinburgh, Castlehill, log-coffins**
NT 2539 7353
NT27SE  121
Lothian Region - City of Edinburgh District
Midlothian - City Parish of Edinburgh

In 1851 two log-coffins were found during the construction of a 'reservoir' on the Castlehill; they were not recovered. The 'reservoir' was probably the water storage tank on the corner of Ramsay Lane, immediately NE of the Castle Esplanade. The excavations reached a depth of 25' (7.6m) at the point where the coffins were found but the depth of the discovery itself is not explicitly stated. They were, however, found in a 'thick layer of moss or decayed animal remains', 'beneath a layer of clay' and 'entirely below the foundations of the ancient city ramparts'.

Each of the two coffins measured about 6' (1.8m) in length and was hollowed out of an oak trunk. They were 'rough and unshapen' externally but had been split open. Internally, they were 'hollowed out with considerable care' to leave shaped recesses for the head and arms in apparent imitation of medieval stone coffins. Each lay E-W, with the head to the W, and contained an unaccompanied inhumation, one of them male and the other female.

The account of the discovery is consistent with their being extempore medieval burials of high status, and possibly connected in some way with military activity at the Castle or with one of the sieges of Edinburgh.

The 'skull and antlers of a gigantic deer', the 'portion of another horn' and the Constantinian coin that were found in close proximity to the burials were presumably derived from earlier contexts.

Wilson 1851, 142, 463-4.

A 21-2  **Edinburgh, Royal Museum of Scotland, 1-2**

In store at the Royal Museum of Scotland there are two trough-type timber artifacts which bear no accession numbers; the possibility that either of them may be a logboat, timber coffin, cooking-trough, mill-trough or similar artifact cannot be discounted. Each of them has been cut to size so as fit their storage rack, in which it is impossible to record them in detail.

1. The better-preserved object measures 2.25m in length. One end is formed by a length of 0.45m left in the solid, while the other has been truncated by sawing. The object is roughly cylindrical (measuring about 0.6m in diameter) and the square-sectioned depression hollowed from the timber measures 0.35m in internal depth. The floor is about 0.2m thick. The timber is in good condition with some splitting at the sawn end, and there are no knots or thickness-gauge holes.

2. The other object measures 2.65m in length between its saw-cut ends. No significant features can be identified and it has split longitudinally into at least two sections in the course of continuing disintegration. *May 1987*

NMRS dataset NT27SE 439.


**Fearnan**  See **Oakbank, crannog, paddle**


**Fishwick Bay**  See **Bailemeonach, Mull, trough**


**Friar's Carse**  See **Carse Loch**

A23  **Glasgow, Bankton, boat**
NS *c.* 571 649
NS56SE  14
Strathclyde Region - City of Glasgow District
Lanarkshire - Govan ph.

The vessel that was found at Bankton in May 1853 was not a logboat in the strict sense, but the unusual nature of its construction, its location within the main area of Clyde logboat discoveries, and its traditional designation as a 'canoe' make it worthy of record.

The boat measured 18' (5.5m) in length, 5' (1.5m) in beam amidships and 3'6" (1.1m) across the stern. A single oak trunk had been dug out to form the keel and the lower part of the hull. Heavy oak frames and overlapping planks had been attached to this section by treenails and possibly also by iron nails, while set into the upper planks there were holes similar to the type commonly drilled for thickness-gauges. The stern was of transom form and at the bow there was a projecting cutwater.

When the boat was discovered (presumably during the construction of Mavisbank Quay) the boat was lying with its keel uppermost and the prow towards the river. It subsequently disintegrated but the remains were transferred to the offices of the River Trustees and have apparently been lost.

Buchanan 1854b; Duncan 1883, 123; [Buchanan] 1884, 353-4.


A24 **Gleann Geal, keg**
NM *c.* 76 53
NM75SE 1
Highland Region - Lochaber District
Argyll - Morvern ph.

In May 1879 a keg of bog butter was discovered in a slanting position at a depth of 4'6'" (1.4m) during peat-digging at the N end of Gleann Geal, Glen Gill or Glen Gell, and possibly in the extensive area of peat-banks around NM 765 537. The keg was donated to the Museum of the Society of Antiquaries of Scotland (now the Royal Museum of Scotland), and is stored under accession number NMS SHC 1 (having formerly been numbered ME 166 and SH 1).

The body of the keg has been worked from a solid trunk or large branch of alder (formerly identified as birch) to form a cylindrical body with slightly barrel-shaped sides. It measures 2' (0.6m) in height and 1'4½" (0.4m) in diameter. Two perforated lugs project to a height of 4" (102mm) above opposite sides of the upper rim and the published illustration also depicts two perforated lugs on opposite sides of the body. Both the lid and bottom of the keg fitted into grooves; the former was 'partly hollowed into a basin shape' and was perforated to allow the lugs to project. The lid (which cannot be identified in the collections) has probably been slightly domed and is illustrated as bearing adze-marks. It had a rectangular slit on one side (to retain one of the lugs), but it is unclear whether there was a slit or a recess on the opposite side.

The keg has recently been the subject of extensive conservation, including the partial epoxy reconstruction of the base, which has evidently been formed from a piece of wood larger than the body which has been so carved as to form a collar which was revealed during conservation to have been retained by small wooden pegs or dowels set into holes in the main body of the object. Internally, the workmanship is generally crude, the bottom of the vessel being uneven and having numerous marks which are probably of natural origin rather than toolmarks. Gouge marks in the exterior have presumably been made at the time of discovery.

The capacity of the vessel was apparently about 75 litres. Bog butter was found completely filling the interior, the upper surface of the mass being domed. Radiocarbon assay of a sample of this material has yielded a determination of 148 ± 35 ad (UB-3185), which determination may be calibrated to between about 183 and 226 cal AD. *November 1994*

Macadam 1882, 220-1; Ritchie 1941, 7, 16; Earwood 1991, 232-3, 234, 235; Earwood 1993a, 109; Crone 1993b, 274; NMS accession register: typescript continuation catalogue.


**Glen Gell** See **Gleann Geal, key**


A25 **Glenfield, Kilmarnock, ard**
NS *c.* 42 36
NS43NW 50
Strathclyde Region - Kilmarnock and Loudoun District
Ayrshire - Kilmarnock ph.

In March 1914 what was identified as a paddle was found at a depth of 5'4" (1.6m) below the surface, and about 191 yards (175m) from the contemporary course of the River Irvine, when gravel was being dug on the land of Messrs. Glenfield and Kennedy, most probably in the area around the present gasworks to the S of Kilmarnock. It was presented to the Dick Institute, Kilmarnock, where it remains in store.

The object (fig. 33) is of wood which has suffered from splitting during drying, and measures 0.86m in length; a roughly-squared shaft of side about 30mm and cut transversely across the end accounts for 0.59m of this. This shaft continues down the centreline of one side of the blade and terminates with a point about 90mm from the end of the blade itself, which is formed as a roughly-equilateral triangle abutting onto a rectangle; the sides curve slightly upwards towards the shafted side. The other surface of the blade is flattened and less regular in shape.

The form of the object differs considerably from that of other Scottish paddles and the asymmetry between the ventral and dorsal surfaces is particularly noticeable. Its size and shape are roughly comparable with that of the two ard-heads that were found during peat-digging at Dale Water, Virdi Field, Dunrossness, Shetland (HU 4035 1917) in 1965 (NMRS dataset HU41NW 10; Rees 1979, i, 43-4), and have been tentatively dated to the Iron Age. The Glenfield 'paddle' was, therefore, more probably an ard-head which (on the basis of its lack of wear) had seen little use.

Munro 1919; Dick Institute, Kilmarnock, museum display label.

OARS, PADDLES AND RELATED ARTIFACTS

Fig. 33  GLENFIELD, Kilmarnock, ard

Fig. 34  LOCH GLASHAN, paddle or oar

Fig. 35  LOCHLEA, crannog, oar

Fig. 36  TENTSMUIR, paddle

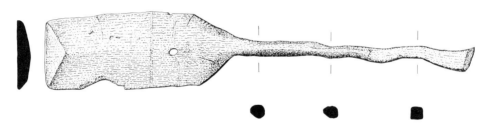

Fig. 37  LOCHLEA, 'double-paddle'

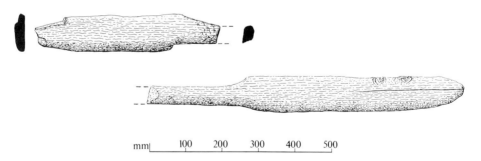

mm  100  200  300  400  500

*Fig. 33. Glenfield, Kilmarnock, ard (no. A25). Plan view with section of blade. Scale 1:10.*
*Fig. 34. Loch Glashan, paddle or oar (possible) (no. A36). Plan of dorsal surface with sections at each end of surviving portion. Scale 1:10.*
*Fig. 35. Lochlea, crannog, oar (no. A51). Plan of blade and part of loom, with sections of distal end of blade and lower part of loom. Scale 1:10.*
*Fig. 36. Tentsmuir, paddle (no. A69). Plan view of dorsal side with sections of distal end of blade, and at three points along loom and handle. Scale 1:10.*
*Fig. 37. Lochlea, crannog, 'double-paddle' (no. A52). Plan view of surviving portion with section of (broken) blade, and at one end of loom. Scale 1:10.*

**Gunnister**  See **Cunnister, Yell, Shetland trough**

### A26  Gutcher, Yell, Shetland, mill-paddle
HU c. 547 989
HU59NW 15
Shetland Islands Area
Zetland - Yell ph.

In store at the Royal Museum of Scotland (under accession number NMS PD 13) there is a probable mill-paddle which was found (probably during peat-digging) at a depth of 7' (2.1m). There was no evidence of a watermill nearby and the local topography was reported to be unsuitable for such an installation.

The object is very light in weight and an attached label identifies the timber as spruce. The leaf-shaped blade is deeply scooped and measures 1'9" (0.3m) in length by up to 6' (152mm) transversely; it varies between 0.85" (22mm) and 2.6" (66mm) in thickness. The handle measures 1¾" (44mm) in length and is evidently not designed for use in the human hand; the large hole and the prominent collar would serve well as attachment-points in a horizontal mill.

This object is most probably a mill-paddle, although its identification as such calls into question the supposition that such artifacts were straight-sided in the far North, as compared to the more elaborately-shaped examples typical of the South-West. *December 1988*

Maxwell 1956, 232.

### A27  Inverlochy, Fort William, timbers
NN c. 112 749
NN17SW 15
Highland Region - Lochaber District
Inverness-shire - Kilmonivaig ph.

In 1986 two timbers were partially exposed by erosion in the S bank of the River Lochy near its entrance into Loch Linnhe. They were uncovered and examined with a view to their being possible logboats, but were found to be heart-rotted and heavily-indurated oak trunks which retained 'many knots and branch stumps'. They bore no evidence of human workmanship.

*Pers. comm.* Mr J. Barber; NMRS MS/438.

**Kilblain, paddle**  See **Kilblain 1**

### A28  Kilmaluag, Skye, keg
NG c. 43 74
NG47SW 9
Highland Region - Skye and Lochalsh District
Inverness-shire - Kilmuir ph.

In 1931 a 'wooden barrel or keg full of...bog-butter' was found in a saturated condition in a peat-bog at Kilmaluag, to the S of the main road and about 1 mile (1.6 km) E of the village. It had apparently been set into bog pool and was in an attitude slightly tilted away from the vertical and about 6' (1.8m) below the level of the former upper surface of the peat. The bottom of the keg was found at the same location in the subsequent year. It was found in a saturated condition and has suffered from warping, splitting and pulverisation due to dessication after discovery; it was taken to the Regional Museum of the Town Council of Aberdeen, but cannot be identified among the collections of the City of Aberdeen, Art and Recreation Division, Art Gallery and Museum, which is the successor to that institution.

The vessel was conserved with glycerine and the exceptionally comprehensive study that was carried out included microscopic, scientific and technical analysis, and formed the basis for a seminal account of this class of artifact. It was assessed to have been lowered into a bog-pool which subsequently became covered-over; pollen analysis of the overlying deposits suggested a date 'Some time in the early historical or later prehistoric periods' but not earlier than late Sub-Boreal or early Sub-Atlantic times. The bog butter contained in the keg has been chemically analysed (with inconclusive results) but neither vessel nor contents has been the subject of radiocarbon dating. Horse, human and cattle hairs have been identified in the butter.

The keg was of birch and comprised body, lid and bottom sections, each of which had been worked from the solid with an adze and a smaller tool. The various holes had been made by burning and the workmanship was of a high standard; neither iron nails nor wooden pegs was used.

The barrel-shaped body measured 1'9" (0.5m) in height and between 3'7" (1.1m) and 3'9" (1.2m) in maximum diameter; the sides were rarely more than ½" (13mm) thick and the capacity was thus about 475 litres. Projecting above the upper rim of the body, and on opposite sides, there were two lugs, each measuring 2¾" (70mm) in height and pierced by two holes, the upper being the larger. A second pair of lugs projected from points a little above the midpoint of the keg and were placed about 1'3" (0.4m) apart on the one side; each was perforated by a 'relatively small' oval hole. Ritchie suggests that the upper lugs were used to lower the keg into place (using ropes passed through the upper holes) while the lower holes were intended to hold a stick to retain the lid; the lower lugs were possibly used to secure the keg as a load on a human porter or pack-animal. The keg had apparently been used on more than one occasion as a major longitudinal drying-split had been repaired with thongs lashed through holes, and a small piece of wood had been so placed as to cover a hole gnawed by mice through the wall.

The circular lid was dressed in a spiral with an adze-like tool, and was slightly convex on the outside but concave within; it was large enough to project slightly beyond the sides of the body, and was retained in place by indentations around the vertical lugs. On the exterior, there was a probable identification mark.

The bottom was formed as a circular base with a bevelled flange, within which the lower edge of the body was fitted; the bottom itself was retained by thongs of leather or hide sewn or laced through corresponding holes in the flange and in the body. Two series of holes around the bottom margin had probably resulted from the replacement of an earlier bottom.

Ritchie 1941; Earwood 1991, 235-6; Earwood 1993a, 109, 110, 157, 159, 161, 163, 219, 277; *pers. comm.*, Ms. J. Stones.

**King Fergus' Isle** See **Loch Laggan 8, boat**

**A29 Kirkbog, paddle**
NX *c.* 872 939
NX89SE 62
Dumfries and Galloway Region - Nithsdale District
Dumfries-shire - Closeburn ph.

In 1862 a 'Canoe Paddle' was found (presumably during drainage operations) on the farm of Kirkbog near Closeburn and 'in a moss lying below a deposit of loam and gravel of about six feet (1.8m) in depth'. Kirkbog farm is situated on the haughland of the middle valley of the Nith at an altitude of about 50m OD, and the overlying deposit was possibly of riverine origin.

The paddle was not recorded in detail and does not appear to survive.

*TDGNHAS*, 1st series, (1865-6), 7.

**A30 Kyleakin, Skye, keg**
NG *c.* 75 26
NG72NE 2
Highland Region - Skye and Lochalsh District
Inverness-shire - Strath ph.

Before 1885 a bronze cauldron of globular form and 'Battersea' type, and 'several kegs or small barrels' of bog butter were found 'in close juxtaposition' during peat-digging near Kyleakin, and at a depth of 7'6" (2.3m). The various discoveries were not recorded in detail and need not have been associated. The cauldron and one of the small kegs were transferred to the museum of the Society of Antiquaries of Scotland (now the Royal Museum of Scotland) where the latter artifact is stored under accession number NMS SHC 2.

This keg was recorded as being hollowed in the shape of a barrel from a single piece of wood. It measured 1'2" (0.4m) in height by up to 1'1" (0.3m) in diameter, making the capacity about 28 litres. Two 'slight projections' or external lugs were noted with holes burnt through them. Both the lid and the base had rested on prepared ledges, and were missing when the discovery was recorded. The timber has been identified as alder and a radiocarbon determination of 220 ± 35 ad (UB-3186) has been obtained from the contents, which may be calibrated to between about 260 and 329 cal AD.

The object was seen in the course of conservation at the Royal Museum of Scotland . It has suffered from extensive splitting and has been reinforced (since discovery) with iron wire around the surviving fill of butter. Neither base nor lid survives but there are the remains of a retaining dowel around the foot and the prominent groove that forms the neck has possibly been intended to hold a retaining string. No evidence of protruding lugs can be identified but two holes which have possibly been burnt through slight projections on each side may have served to retain strings. *November 1994*

Anderson 1885, 309-11; Ritchie 1941, 15, 16, 20-1; Macgregor 1976, 2, no. 306; Earwood 1991, 233, 235, 236, 237; Earwood 1993a, 13, 14, 109-11, 278.

**A31 Landis, timbers**
NX 975 663
NX96NE 12
Dumfries and Galloway Region - Nithsdale District
Kirkcudbrightshire - New Abbey ph.

The discovery of 'lengths of dressed bog-timber' has been reported about 400m SE of Landis farmhouse in the haughland of the lower Nith, and at an altitude of 10m OD. The timbers were reported to be cut into standard lengths of about 2.5m; some of them bore toolmarks.

*DES*, (1956), 17.

**A32 Loch a' Ghlinne Bhig, Skye, bowl**
NG *c.* 41 44
NG44SW 3
Highland Region - Skye and Lochalsh District
Inverness-shire - Bracadale ph.

This bowl was apparently discovered between 1979 and 1981 when a forestry road was being cut in the area between Loch a' Ghlinne Bhig and the River Snizort, and to the N of the B885 public road; its precise findspot, context and contents (if any) were apparently not recorded but it was evidently found in rough ground and not associated with other archaeological remains. It is now held by the Museums Service of Skye and Lochalsh District

Council, and is of particular interest on account of its being unfinished and thus indicating the mode of manufacture.

In form, the vessel is roughly comparable with that from Talisker Moor (no. A68), being roughly circular, shallow, round-bottomed and markedly carinated. A short carved stud projects from the exterior at the level of the carination and directly opposite the prominent lug that has apparently served as a handle. On the evidence of the published drawings, the capacity may be estimated at about 2 litres in its unfinished state.

The vessel was carved from a knot-ridden section of alder tree measuring about 0.2m in diameter, the grain of the wood lying roughly parallel to the plane of the rim and the centre of the tree passing through the bowl at the level of the carination. The eccentricity of the vessel may have been governed by the presence of the numerous knots, but may also be due to its having been allowed to dry out before transfer to the museum.

The bottom of the vessel is smooth and devoid of toolmarks but the oval compression-mark that encircles the lower part may indicate the former position of a metal ring (which may either have been used to support the vessel during manufacture, or been added after discovery. Shallow grooves below (and parallel to) the carination cover the greater part of the lower vessel; the ridges have been left unsmoothed, possibly as a deliberate part of the design. Cut-marks around the junction of the lug with the bowl indicate the process of manufacture, while the neck has been shaped with a series of shallow and vertically-aligned facets. The variability in workmanship around the (flat) rim and the upper part of the vessel suggests that it remains unfinished. The interior clearly requires further work, and bears a rosette of rough and sharp toolmarks around which there survive splinters, strongly suggesting that the vessel has never been used.

Crone interprets these marks as suggesting the following sequence of manufacture:

1. initial shaping (with an axe) from which no marks remain,

2. careful external shaping, probably with a shallow gouge, 10-12mm broad at the cutting edge,

3. interior dug out, probably with a small axe, about 40mm broad at the cutting edge.

Crone places the bowl within the group of handled bowls and vessels that have been worked by either carving or lathe-turning. Unusually, it may have been lidded, the lid being retained by a strap secured around the stud.

The vessel has been radiocarbon-dated to 20 ± 50 ad (UT-1698), which determination may be calibrated to between about 52 and 63 cal AD. As such, it is broadly contemporary with that from Talisker, and may be a product of the same workshop. Both are seen as functionally distinct from the flat-bottomed pottery forms that appear locally contemporary, and may have been intended to be portable, possibly slung over the shoulder or suspended from a belt. Both may be placed within Earwood's Armagh group.

Earwood 1993a, 267; Crone 1993b.

**Loch Carron**  See **Plockton. keg**

A33  **Loch Coille-Bharr, paddle**
NR c. 779  894
NR78NE  14
Strathclyde Region - Argyll and Bute District
Argyll - North Knapdale ph.

In 1887 underwater investigation revealed a paddle under about 6' (1.8m) depth of mud to the E of the crannog in the SE corner of Loch Coille-Bharr (or Kielziebar) which is situated in the craggy land of Knapdale at an altitude of about 35m OD.

The paddle (which has apparently been lost) was of 'oak' and was 'elegantly made, like a barbed arrow'. It was described as concave on one side but convex on the other. The point of the blade was missing and the handle, which was about 4' (1.2m) long, became detached and was lost soon after its discovery.

The blade measured 8½" (216mm) in length, 8" (203mm) in greatest width, 2" (51mm) in width 'at the point' and 2" (51mm) in thickness, as discovered. The surviving portion of the handle measured 2" (51mm) in diameter and in it there were two blind holes measuring 2" (51mm) in diameter and set 3" (76mm) apart 'as if pegs had been inserted'.

Mapleton 1868, 323; Campbell and Sandeman 1962, 125, no. 85.

A34  **Loch Doon, boat fragments**
NX 486  940
NX49SE  12
Strathclyde Region - Cumnock and Doon Valley District
Ayrshire - Straiton ph.

In 1972 artifacts including the thwart-board of a boat and an oak thole-pin were found by sports divers 'off-shore of Castle Island'.

*DES*, (1972), 12.

**Loch Doon, paddle**  See **Loch Doon 1-6**

### A35 Loch Eport, North Uist, trough
Western Isles Area
Inverness-shire - North Uist ph.

In June 1895 an open-topped dish or trough which was found during the construction of a road through 'a rising ground of solid peat' along the side of Loch Eport and 'a little distance to the east of the kelp works'; this provenance cannot be precisely identified. It lay in a slanting attitude at a depth of five or six feet (1.5-1.8m); there was a similar depth of peat below the object, which MacRitchie believed to have sunk from the surface when the peat was in a more liquid condition. The object was 'quite empty' when discovered and apparently worked from the root portion of a Scots fir. The object has evidently been formerly accessed as 1896.375 and 1954-692 and is described on the museum label as a 'dish or butter boat'.

At the time of discovery, MacRitchie cited the principal external dimensions as being 2'4" (0.71m) in length and 11" (0.28m) in breadth. The depth of the hollow was 5" (0.13m) and the internal volume may thus be tentatively estimated at about 1.4 litres. The 'underpart of each extremity' had been fashioned to form a 'rudely shaped hollow' which was 'evidently for lifting'.

In store at the Royal Museum of Scotland (under accession number NMS ME 993) there is an object which bears a (loose) label describing it as this object, but is more probably to be identified as that from Cunnister (no. A13).

MacRitchie 1896; Earwood 1993b, 357; NMRS dataset NF86SE 10.

### A36 Loch Glashan, paddle or oar (possible)
NR c. 920 934
NR99SW 23
Strathclyde Region - Argyll and Bute District
Argyll - Kilmichael Glassary ph.

The 'possible wooden paddle or oar' (fig. 34) that was found close to the Loch Glashan 1 logboat (no. 102) on the E shore of the loch (and may have been associated with it) was not conserved, but at least part survives in the store of the Glasgow Art Gallery and Museum, where the storage label for accession number GAGM A6137b identifies 'Part of paddle (?) found near canoe at Loch Glashan'.

The piece of wood is light in both colour and weight, and measures 0.41m in length by up to 68mm in breadth and 15mm in thickness. It has suffered from flaking during drying-out, and has been snapped at the proximal end, where it has evidently tapered into a shaft or handle. The distal end tapers to a ridge in the longitudinal plane and the ventral surface is slightly concave in all three planes.

*August 1989*

### A37-44 Loch Glashan, crannog
NR 9159 9249
NR99SW 1
Strathclyde Region - Argyll and Bute District
Argyll - Kilmichael Glassary ph.

The unpublished excavation of 1960 upon the Loch Glashan crannog that is noted under no. 103 revealed that the upper substructure comprised oak and silver birch logs laid over a brushwood layer, in which there was found a wide range of artifacts, including sufficient timber objects to indicate both the wide variety of functions and techniques employed in domestic woodworking at this period, and the quantity of material that may be expected from the excavation of such monuments. A further parallel (nos. A50-3) may be noted at Lochlea crannog.

The heavily-waterlogged and uneroded nature of this layer accounts for the survival of much organic material, including both wood and leather. Following recognition of the multi-period nature of crannogs (Guido 1974; Lynn 1986) and the elucidation of the stratigraphic series at Buston (nos. 11-12), Earwood must be considered correct in proposing that the layer does not represent the foundation or substructure, but was formed by periodically raising the height of the crannog above one or more occupation-surfaces. The excavations at Loch Glashan did not penetrate below the (then) water level.

The occupation of the crannog has been attributed in general terms to between the 6th and 8th centuries AD, on the basis of the pottery recovered. Earwood (1990, 85) notes that the form of one (fragmentary) turned wooden vessel closely parallels that of E-ware pottery.

Following the publication of a comprehensive account (Earwood 1990) a descriptive and quantitative summary of the recorded examples of oars, paddles and monoxylous troughs and other artifacts germane to this study is given to indicate both the variety of forms and techniques, and the relative frequency of their manufacture by monoxylous reduction. All the surviving finds (which have been extensively conserved and restored) are in store at Glasgow Art Gallery and Museum. The following representative examples are of particular interest:

1. (A37) This 'Oar-shaped implement' was found with the blade end about 1' (0.3m) deep in the brushwood layer and the other end just projecting above the surface of that layer. It was found in a slanting attitude and upper end was 'weathered'. It was allocated small find no. 40 upon discovery and is in store under accession number GAGM A6046bn.

The object is of rectangular section with rounded corners, and has suffered slightly from splitting. It measures 0.25m in length over all, and 30mm in thickness, and varies in breadth between 98mm at the broadest point and only 40mm at the narrow end, where it is broken. The widened and flattened broader end has been pierced by a now-incomplete hole which has probably

*Pl. 20. Loch Glashan, crannog, paddle (no. A38). Photograph taken* in situ *during excavation and reproduced from the negative held in the NMRS (A53507) by kind permission of Mr. J.G. Scott and of the Royal Commission on the Ancient and Historical Monuments of Scotland.*

been oval, having axial measurements of about 45mm and 60mm.

Although this fragment bears a superficial resemblance to the transitional section of an oar or paddle between the blade and the shaft, it has more probably formed part of a structural member; the well-formed hole would have formed one element of a joint. *August 1989*

2. (A38) This object (pl. 20) was apparently found complete at a depth of 1' or 2' (25mm or 51mm) in the brushwood layer. It was allocated small finds number 8 and identified as a wooden paddle, which identification appears correct. It is in store under accession number GAGM A6046bl. The wood has been rendered dark black during conservation treatment (probably with Carbowax) and numerous breaks have been repaired and infilled, but the object remains highly fragile, so that only the dorsal surface could be examined at the date of visit.

The object measures 1.1m in length by up to 0.1m transversely, and is curved towards the ventral surface and to one side. The shape is only roughly-formed so that the junction of the blade with the handle is asymmetrically-rounded in form. The handle itself is roughly circular in section and measures about 40mm in diameter along most of its length; it becomes more pointed towards the end,

where it has been cut square across. The timber of the handle is heavily-knotted, and into it there are cut two holes of unknown origin and a depression which has possibly been worked with a tool.

The blade is severely chipped around the sides and at the end, and has possibly lost some of its length. On the examined side there is a slight midrib, which measures about 15mm in thickness and tapers towards the edge. *October 1987*

NMRS A53507.

3. (A39) This 'paddle-like object' (fig. 39) was found 'set almost upright' in the brushwood layer; it was 'broken, with central hole'. Following numbering as small find no. 28, it was conserved and placed in store under accession number GAGM A6046bm.

Two fragments of conserved timber that are held under that number fit together at the break and have probably formed part of a paddle of relatively sturdy construction. Neither of them is pierced, which may suggest that the object is incompletely preserved.

One piece has a roughly rectangular section measuring about 50mm by 45mm, is about 0.26m long and apparently formed the lower part of the shaft of the handle.

*Pl. 21. Loch Glashan, crannog, paddle-shaped object or 'bat' (no. A40). Studio photograph with another timber artefact, the 'bat' being the lower. Reproduced from the negative held in the NMRS (A36978) by kind permission of Mr. J.G. Scott and of the Royal Commission on the Ancient and Historical Monuments of Scotland.*

This piece has broken away at the point where it broadens into what was apparently a spade-type grip at the proximal end. The fragment measures 26.5mm in length and comprises the tapered junction of the shaft with the handle. The blade has apparently measured 90mm in breadth by 25mm in thickness and been roughly rectangular with rounded corners, a flat (presumably ventral) surface and a slightly convex (presumably dorsal) side. *August 1989*

4. (A40) The 'paddle-blade' or 'bat' that was found at a depth of about 2' (0.6m) in the brushwood layer was numbered as small find 119 and is stored under accession number GAGM A6046az. It has suffered slightly from splitting, and has been mounted (after conservation) on a board so that only one surface is available for inspection. Although this object cannot be specifically identified, it appears totally unsuitable for boat propulsion, and most probably had a domestic function, possibly as a platter or skillet. Earwood considers a variety of ethnographic parallels for this object and suggests a tentative identification as a flax beater, which would be consistent with one proposed function of A41 (below).

The object (pl. 21, lower left) is of flat form, and measures 0.43m in length over all by 15mm in thickness. The pear-shaped blade is 260mm long and is now some 160mm broad but has been worn asymmetrically and was probably originally some 50mm broader; part of one corner has been chipped away. The handle is clearly differentiated and measures 105mm in length by 30mm in width; at the proximal end there is an enlarged circular section which measures 55mm in diameter and is pierced by a central hole of diameter 11m. All the junctions between surfaces are rounded, except the edges of the central hole which are more sharply defined; the blade was apparently split in antiquity. *August 1989*

NMRS A36980 and A53486.

5. (A41) Listed as excavation finds number 57, there is an object (fig. 38) of indeterminate function which has the form of a miniature paddle. It was found at a depth of

## III Examples of Related Artefact-types

## OARS, PADDLES AND RELATED ARTIFACTS

Fig. 38  LOCH GLASHAN, crannog, possible model paddle

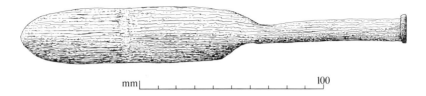

Fig. 39  LOCH GLASHAN, crannog, paddle

Fig. 40  LOCH KINORD, paddle

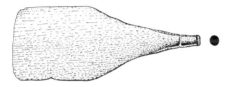

Fig. 41  RAVENSTONE MOSS, paddle

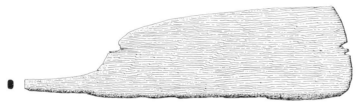

Fig. 42  RUBH' AN DUNAIN, possible oar or paddle

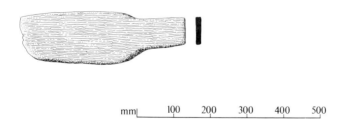

*Fig. 38. Loch Glashan, crannog, model paddle (possible) (no. A41). Plan view at scale 1:2.*
*Fig. 39. Loch Glashan, crannog, paddle (no. A39). Plan of dorsal side. Scale 1:10.*
*Fig. 40. Loch Kinford, paddle (no. A46). Plan view of surviving portion, with section of lower loom. Scale 1:10.*
*Fig. 41. Ravenstone Moss, paddle (no. A62). Plan view of surviving portion with section of lower loom. Scale 1:10.*
*Fig. 42. Rubh' an Dunain, Skye, oar or paddle (possible) (no. A65). Plan view of surviving portion, with section of lower loom. Scale 1:10.*

1'9" (0.5m) in the brushwood layer, and is in store under accession number GAGM A6046bb. It had broken into three pieces, which were re-united during conservation with Carbowax, but the object remains fragile and could not be handled at the date of visit.

It has been worked from close-grained wood and measures 8 3/16" (208mm) in length over all. The blade is of an elongated-tongue shape with a rounded end, and measures up to 60mm in width by about 3/16" (4.8mm) in thickness; it is almost flat on one side, has a slight 'keel' on the other, and passes gradually into the handle, which is oval in section with a distinct thickened terminal.

The finds list classifies it as a weaver's (or sword) beater, but it more probably had a domestic function or was, just possibly, a toy or model paddle, possibly with a similar ceremonial or ritual function to that of the numerous model religious objects found in Romano-British contexts. Earwood identifies it as a probable spatula (considering it too short for a weaver's beater) and catalogues several other objects of similar type from Loch Glashan besides citing parallels from the excavated Irish crannogs of Ballinderry 2 and Lagore. Wild confirms that the small size and apparent light weight of this object would render it of no use except in the weaving of tapestry, a craft not normally associated with crannogs.

The discovery of several spatula-blades is also recorded; these were probably for culinary use. *October 1987*

6. (A42) The object (GAGM A6046bq) that is numbered 55 in the finds list was classified at the time of discovery as a 'wooden bowl', but may, however, be better considered as a trough of the same general form as those from Cnoc Leathann, Durness (no. A11) and Eadarloch (no. A18) with square ends and ledge handles. It is the largest wooden vessel from the crannog, and Earwood cites an additional close parallel in an Irish Early Christian context at Deer Park Farm, Co. Antrim.

It was found 'almost upside down' and one side had been crushed by a large squared timber, while evidence of worm attack is visible in places. It has acquired a reddish-brown colour during treatment, but remains fragile so that the bottom could not be examined at the date of visit.

The trough (fig. 44) has apparently been worked from a whole log measuring about 0.3m in breadth, and subsequently rubbed or sanded down to remove any toolmarks. It measures 0.94m in length over all by up to 0.28m transversely and 135mm in external depth, and has probably been rectangular in form, but one side has become bowed outwards from the pressure of overlying deposits whilst waterlogged. It is probable that the vessel formerly measured about 0.21m in breadth. The sides measure about 90mm in thickness and the ends slope inwards both internally and externally, while the full-length ledge handles measure about 100mm in length along the long axis of the trough and were probably originally of playing-card form, but have become chipped and rendered irregular by post-depositional abrasion. Externally, the sides meet the bottom in a succession of slight steps.

The internal cavity has measured about 0.78m and 0.65m in length at the top and bottom respectively by 0.21m in breadth and 0.12m in depth, which equates to a capacity of about 18 litres. *February 1989*

7. (A43) The trough numbered 120 in the finds list is of broadly similar form to the larger vessel (no. A42) across which it was found lying at a depth of 1'10" (0.6m) in the brushwood layer. The central section of the base and most of one end have been lost through splitting (apparently in antiquity); further (subsequent) splitting and wear are also evident. Following conservation, it has been placed in store under accession number GAGM A6046bo.

This ledge-ended trough or squared bowl (fig. 45) is described in the finds list as small and 'square-ended'. Externally, it measures 0.43m in length over all by 0.23m transversely and 90mm in depth; the corresponding internal measurements are 0.34m, 0.2m and 65mm respectively, and the flat bottom measures 285mm by 150mm. The capacity is about 4.4 litres. *August 1989*

NMRS A53806 and A53808.

8. (A44) In store under accession number GAGM A6046bp and classified as a trough, there is the object that is numbered 121 in the finds list and is there considered as a domestic bowl or dish, being captioned as such in the illustrations to the present work (pl. 22, upper and fig. 46) but is more probably to be classified as a small trough. It measures 0.7m in length over its rounded ends and projecting handles by up to 0.29m transversely and 125mm in depth. The flat bottom measures up to 0.5m by 150mm internally, and the capacity is about 12 litres. The tree-ring pattern is obscured but the grain runs from end to end and there is some evidence of splitting.

The body of the bowl is rounded in section but flat-bottomed, and may loosely be described as pear-shaped. The larger of the two handles is located roughly one-quarter the distance down the side at the higher of the two ends, and within the interior there are the faint marks of a chisel, gouge or similar instrument. *February 1989*

RCAHMS 1988, 35, 205-8, no. 354; Earwood 1990; Earwood 1993a, 95-6, 97, 98, 100, 101, 102, 103, 107, 118, 119, 121, 122, 123, 130, 131, 137, 148, 164, 223, 224, 280; MS. notes in GAGM: excavation finds list; *pers. comm.* Dr JP Wild.

**Loch Kielziebar** See **Loch Coille-Bharr**

**A45 Loch Kinord 5, boat**
NO *c.* 44 99

TROUGHS AND BOWLS

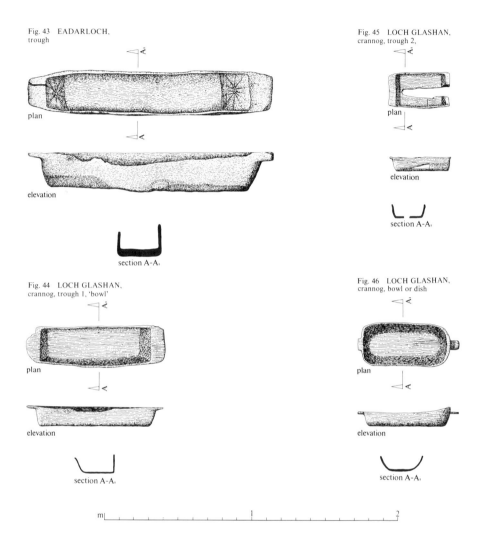

Fig. 43. Eadarloch, trough (no. A18). Plan and elevation with section at miidpoint. Scale 1:25.
Fig. 44. Loch Glashan, crannog, trough 1 (no. A42). Plan and elevation with section (at midpoint). Scale 1:25.
Fig. 45. Loch Glashan, crannog, trough 2 (no. A43). Plan and elevation with section (at midpoint). Scale 1:25.
Fig. 46. Loch Glashan, crannog, trough 3 (no. A44). Plan and elevation with section (at midpoint). Scale 1:25.

NO49NW 33
Grampian Region - Kincardine and Deeside District
Aberdeenshire - Glenmuick, Tullich and Glengairn ph.

In about 1845 a boat was discovered in the mud of the loch-bed and brought to shore. It was 26' (7.9m) long and the flat bottom measured 8' (2.5m) in beam. The form of the boat was 'Sharp at both ends, otherwise coble built' and it was 'substantially built' of caulked oak planks secured by 'iron bolts well made and skilfully rivetted'. It was not a logboat.

The location of the discovery is not accurately recorded but Michie attributes it to the last period of occupation of the castle, in the early 17th century. He also mentions the frequent discovery of boat-planking in the waters of the loch.

Stuart 1866, 171; Michie 1910, 83-4.

**A46  Loch Kinord, paddle**
NO c. 44 99
NO49NW  34
Grampian Region - Kincardine and Deeside District
Aberdeenshire - Glenmuick, Tullich and Glengairn ph.

In store at Aberdeen University Anthropological Museum (under accession number AUAM 576-2) there is a paddle-

*Pl. 22. Loch Glashan, crannog, bowls and dishes, including no. A44: upper. Studio photograph reproduced from the negative held in the NMRS (A36975) by kind permission of Mr. J.G. Scott and of the Royal Commission on the Ancient and Historical Monuments of Scotland.*

blade which was found in a 'lake-dwelling' in Loch Kinord. This loch is situated in moorland to the N of the Dee valley at an altitude of about 165m OD; a castle is recorded on an island at NO 4397 9964 and a crannog at NO 4433 9952.

The artifact (fig. 40) is in good condition and only slightly split. It measures just over 0.52m in length, of which the near-rectangular section of the blade accounts for 0.22m and the concave shoulders for a further 0.24m. The blade measures about 12mm in thickness and the dorsal surface is very slightly rounded, while the ventral surface is flat and measures up to 0.2m in breadth. The handle is slightly oval in section and measures about 20mm in diameter; it has broken away at a point where a slight tongue of wood protrudes (possibly as part of a socket joint), and about 50mm lower down the handle there is a slight groove of undetermined function. *September 1987*

*Pers. comm.* Mr. J. Inglis; MS. note in AUAM; NMRS datasets NO49NW 16 and 17.

**A47 Loch Laggan, bowl**
NN *c.* 535 895
NN58NW 6
Highland Region - Badenoch and Strathspey District
Inverness-shire - Laggan ph.

The bowl that was found in the Loch Laggan 2 logboat (no. 110) has become severely warped and split on drying into seven pieces which are held in the Royal Museum of Scotland under accession numbers NMS HX 276:1-7 respectively. It has a narrow everted rim and has been tentatively reconstructed (by Henshall) as relatively shallow, round-bottomed and circular in form, measuring about 5½" (14mm) in diameter. She noted a pierced vertical handle on one side and supposed there to have been a second opposite.

Earwood sees the object as comparable to one (NMS ME 65) of those from Ardgour (nos. A1-4), suggests that it has been turned from one piece of birch and attributes it to the early or mid first millennium AD.

Maxwell 1951, 163-4; Earwood 1990b, 39; Earwood 1993a, 26, 64, 65, 280.

### A48  Loch Laggan 8, boat
NN 4987 8755
NN48NE 4
Highland Region - Badenoch and Strathspey District
Inverness-shire - Laggan ph.

In 1934 an 'old built boat (now in pieces)' was found on King Fergus' Isle. Four fragments of it are held in the Royal Museum of Scotland under accession numbers NMS HX 281, 1-4; three of them have been identified as a patch, a floor-timber and a length of two conjoined sections of plank. In the repair-patch, there was found the head of a hand-made nail and between the two sections of plank there was hair-derived caulking-material.

Maxwell 1951, 162-3.

**Loch nan Eala**  See **Arisaig, Loch nan Eala**

**Loch Spynie**  See **Easter Oakenhead**

**Loch Treig**  See **Eadarloch**

### A49  Lochar Moss, shoulder-yoke
NY c. 00 78
NY07NW 33
Dumfries and Galloway Region - Nithsdale District
Dumfriesshire - Tinwald ph.

On display in Dumfries Museum (under accession number DUMFM 1936.5) there is an object which was formerly said to be a double-bladed paddle but has been re-identified as a shoulder-yoke.

*Pers. comm.* Mr. D. Lockwood.

### A50  Lochlea (crannog) 6, 'boat'
NS 4574 3026
NS43SE 5
Strathclyde Region - Kyle and Carrick District
Ayrshire - Tarbolton ph.

In 1878-9 excavation of the crannog at Lochlea or Lochlee revealed what were identified as parts of a boat. Although not a logboat, the situation of this discovery and its proximity to logboats Lochlea 1-5 (nos. 124-8) make it worthy of record.

The fragments (which are now lost) comprised two 'very curious beams' situated 7'9" (2.4m) apart, lying above the log pavement and separated from it by a thin clay layer. Each of them had a raised 'rim' along its length, a horizontal hole to take the end of a beam, and square-cut holes which were apparently intended to retain uprights; one of these holes was slightly curved. The discovery of a 'paddle' and 'oars' (nos. A51-2) nearby may be coincidental.

In the absence of a complete description, it is impossible either to determine the type and characteristics of the boat, or to exclude the possibility that the two 'beams' should more correctly have been identified as the basal supports for a timber building.

Munro 1880, 39.

### A51  Lochlea, crannog, oars
NS 4574 3026
NS43SE 5
Strathclyde Region - Kyle and Carrick District
Ayrshire - Tarbolton ph.

Munro notes the discovery of a 'large oar, together with the blade portion of another' during the excavation of the Lochlea crannog. On display at the Dick Institute, Kilmarnock, there is a wooden object (fig. 35) which may be one of these finds. It measures 1.04m in length over all and comprises a blade which is bilaterally symmetrical and straight-sided (measuring about 0.86m in length, 80mm in breadth and 30mm in thickness) with asymmetrical sloping shoulders and what is presumably the stump of a shaft of a handle of irregularly-rounded section, measuring about 40mm in diameter. *December 1987*
Munro 1880, 67.

### A52  Lochlea, crannog, 'double-paddle'
NS 4574 3026
NS43SE 5
Strathclyde Region - Kyle and Carrick District
Ayrshire - Tarbolton ph.

The object (fig. 37) that Munro depicts and identifies as a paddle is displayed in the Dick Institute, Kilmarnock, under accession number KIMMG:AR/D 33; it was originally accessed as LC or GW 243. It has broken into two pieces across the narrowest point and has suffered from chipping and splitting in parts.

When reconstructed, it measures about 1.37m in length and comprises two roughly-oval blades joined by a shaft of roughly-circular section which measures about 30mm in thickness. The two blades are about 30mm thick and measure roughly 0.42m by 0.1m and 0.6m by 90mm respectively; the shorter is chipped around the end and the two were probably originally of comparable size.

Although this object bears a superficial resemblance to a double-bladed paddle, it is unparalleled in Scotland and appears too short for practical use so that the suggested identification cannot be accepted with certainty. It is probably functionally (if not necessarily chronologically) comparable to the double paddle-spades that Lerche notes from contexts in the Danish Iron Age; there is also a single Swedish example. Some 280 of these objects have been found (most commonly during peat-digging) in north-central Jutland, and traditionally identified as double-oars related to the numerous logboat discoveries in the region; the example from Stokholm bog was found close to such a vessel. However, the frequency of their discovery with peat-digging spades and in a worn and chipped condition, has caused them to be re-assessed as double-spades on the basis of ethnographic parallels.

Two of the Danish examples have been dated by radiocarbon to the Celtic Iron Age; those from Rybjerg bog and Holmegard bog yielded dates of 330 ± 100 and 220 ± 100 bc respectively. All are of monoxylous form, most being of oak. Typically they measure between 0.9 and 1.6m in length over all, the blades being between 0.27 and 0.84m long by between 80 and 110mm broad, and between 20 and 40mm thick. The blades of the recorded Danish examples vary considerably in cross-section but are apparently typically asymmetrical, the junction between the blade and the handle being generally distinct. The Lochlea artifact shares all these characteristics.

Within the Danish examples, there is an apparent division between two sub-types which are differentiated by the length of the shaft between the blades, this being either of the order of 1½ to 2, or between 3 and 5 handspans. The Lochlea example apparently falls into the second category.

There is no indication that the Lochlea double-spade shares a chronological or 'cultural' link with the Danish examples, but the morphological and functional similarities place it in the same class. *December 1987*

Munro 1880, 67; Lerche 1977.


**A53 Lochlea, crannog, trough**
NS 4574 3026
NS43SE 5
Strathclyde Region - Kyle and Carrick District
Ayrshire - Tarbolton ph.

Munro's excavations on the Lochlea (or Lochlee) crannog revealed a 'beautiful trough cut out of a single block of wood'. It had evidently been wrought with a 'gouge-like instrument' from a block of soft wood, and disintegrated soon after its discovery. It was apparently found at a depth of 5' (1.5m) in the crannog structure, between the margin and the circle of stakes surrounding the log pavement.

On the evidence of the published figure, this object measured about 1m in length over projecting full-width horizontal or ledge-projections, and about 0.6m internally. The breadth was probably about 0.4m and the object appears to have measured about 0.3m in depth. The sides and bottom were thin, and had suffered serious damage. The cavity was rectangular in all three planes, and the capacity may be calculated at about 70 litres.

Munro 1880, 45-6; Earwood 1990, 84; Earwood 1993a, 36, 94, 100, 102, 128, 280-1.


**Lochlundie Moss, oars**  See **Lochlundie Moss**


**Loch nan Eala**  See **Arisaig, Loch nan Eala**


**Longside**  See **Cairngall**


**A54 Midtown, trough**
NG *c.* 820 856
NG88NW 1
Highland Region - Ross and Cromarty District
Ross and Cromarty - Gairloch ph.

In May 1893 a 'dish' or trough was discovered when peat was being dug about 200 yards (182m) N of the former schoolhouse at Old Midton and about 10 yards (9.1m) W of the B8057 public road. The area is situated near the W shore of Loch Ewe at an altitude of about 25m OD, and the object had probably been covered by at least 3m depth of peat deposit. Part of one end was lost during recovery and the fragment of 'straw or hay rope' that was found beside the trough disintegrated upon exposure to air. The fate of the trough itself is unknown.

On examination the object was found to be worked from a single piece of what was probably well-preserved bog-oak. Its form was not recorded in detail and no drawings were made, but it was said to be oblong on plan and the upper parts of each end projected for about 1½" (38mm) to form what were probably ledge-handles; one of these had been pierced vertically as if the trough were to be stored hanging. There was a second hole, which had been plugged, in the bottom at the same end.

The recorded external measurements were 2'1" (0.6m) in length by 1'2" (0.4m) transversely and 4" (0.1m) in depth, and the corresponding internal dimensions were 1'8" (0.5m), 1'1" (0.3m) and 4" (0.1m) respectively. The sides and bottom measured between ¼" (6mm) and ½" (13mm) in thickness, and the bottom of the cavity measured 1'6" (0.5m) by 11" (0.3m).

On the basis of these figures and on the assumption that it was roughly rectangular, the volume of the cavity was about 15 litres. When found, it was filled with bog butter which had evidently been poured into it in a molten state.

Macrae 1894; Earwood 1993a, 100, 102, 283.

**Mill of Williamston**  See **Williamston**


**Milton Island, paddle (possible)**  See **Milton Island**


**Ness of Sound**  See **Staura Cottage, Shetland, paddle**


**A55  Newstead, steering-oar**
NT 570 344
NT53SE 20
Borders Region - Ettrick and Lauderdale District
Roxburghshire - Melrose ph.

The Roman fort of Newstead and its attendant annexes and temporary camps form one of the large complexes of Roman military remains in the Borders, but there are few surface remains to be seen at the site, which is situated on the S haughland of the River Tweed at an altitude of about 120m OD and some 40m above the river. Occupation of various elements of the complex has been demonstrated for the Flavian (Agricolan and Domitianic), Antonine-Trajanic and Severan periods, while the importance of the complex is further increased by the discovery of organic remains in an exceptional state of preservation during excavations between 1905 and 1910.

Within the complex there were found numerous shafts, pits and wells which have long attracted attention on account of the possible Celtic ritual significance that is suggested by the discovery within them of animal and human remains (including severed heads), broken pottery, weapons and metalwork of high value, and a discarded Roman altar. Pit LXV, which was discovered to the N of the fort at NT *c.* 5694 3458 was excavated in 1908 and found to measure 17' (5.2m) in depth; it was rectangular on plan and tapered between 7'6" (2.3m) by 6' (1.8m) at the top and 4'3" (1.3m) by 3'6" (1.1m) at the bottom. The lower 8' (2.4m) of the pit were filled with a blackearth deposit within which were found fragments of metalwork (including a fibula-spring, a surgeon's probe and a lamp), fragments of pottery (including black burnished and Samian wares), a possible wheel-spoke and a steering-oar. Coins of the republican Gens Cordia (46 BC) and of Galba (68 AD) were found at the bottom and above the top respectively of the blackearth, which is tentatively dated to the later first century AD.

Following restoration, the steering-oar is displayed in the Royal Museum of Scotland under accession number NMS FRA 1131. It is inaccessible in a case but its principal measurements were recorded by Curle at the time of discovery. He noted that it measured 5'5" (1.65m) in length. The blade has the form of a rounded and asymmetrical near-rectangle about 5½" (140mm) wide, and appears to account for about 40% of the overall length. The asymmetrical shoulders lead into the loom which measures about 7" (178mm) in circumference, and the timber has been identified as oak.

The oar is pierced transversely across its broader axis by two holes which serve to identify it as a steering-oar, rather than one used for propulsion. That cut 5" (127mm) from the end of the loom is 1¼" (32mm) square and was probably intended to retain an inwards-pointing tiller, while that cut at the lower end of the blade probably served to retain a lanyard which partly supported the weight of the oar and was possibly also used to raise it during beaching operations. No traces are visible of the collar that must have held the oar at its mid-point. *April 1989*

Curle 1911, 131, 313, pl. LXIX.5; Marsden 1963; Marsden [1967], 33; Ross 1968, 269, 283; Ross and Feachem 1976; Clarke, Breeze and Mackay 1980, 69, no. 76; Hanson and Maxwell 1983, *passim*; McGrail 1987a, 242, 243-4; Marsden 1994, 75.


**A56  Oakbank, crannog, paddle**
NN 7230 4429
NN74SW 16
Tayside Region - Perth and Kinross District
Perthshire - Kenmore ph.

Oakbank crannog is situated near the N shore of Loch Tay which is situated in a deep glaciated trough in the Trossachs at an altitude of 107m OD. It lies in front of Oakbank cottage and 190m ESE of Fearnan Post Office; the site-name Fearnan has been attached both to this crannog and to that (otherwise known as Fearnan Hotel) situated some 220m to the W.

Underwater survey has shown that the main crannog-mound measures about 25m in maximum diameter. The highest point is about 1m below the normal level of the loch and the bottom of the mound is at a depth of about 2m on the NW side and about 3.6m on the SE. On the SW there is a projecting 'smaller circular feature' (possibly the remains of a landing-stage) measuring about 7m in diameter and linked to the main mound by a scatter of boulders. The radiocarbon testing of piles and timbers has yielded four dates ranging between 595 ± 55 bc (GU-1323) and 410 ± 60 bc (GU-1463), which may be calibrated to between 788 and 697, and about 404 cal BC respectively.

In 1981 a paddle (pl. 23) was found during excavation of the organic matrix of the crannog. It is currently undergoing restoration and conservation but is reported to measure about 1.35m in length, being 'shaped very like a modern paddle with a rounded, slightly plano-convex cross-section'. The published photographs show that the blade is roughly rectangular in form and occupies about 52% of the surviving length. The junction between the blade and the handle is protracted and the latter feature (much of which has broken away) is apparently of rounded section. The timber has been identified as oak and bears numerous clear toolmarks.

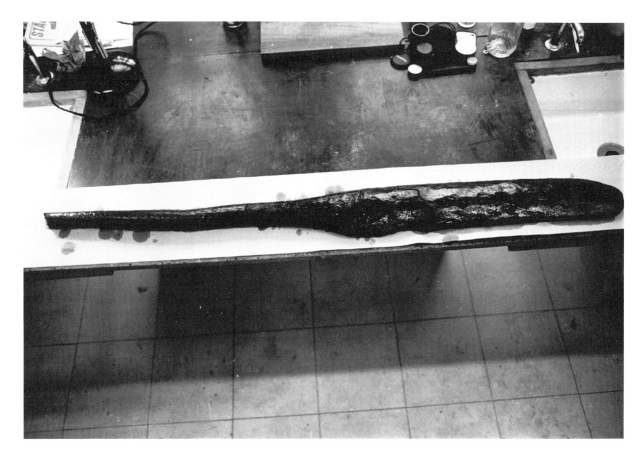

*Pl. 23. Oakbank, crannog, paddle (no. A56). Dorsal view reproduced by kind permission of Dr. T.N. Dixon.*

Dixon 1981; Dixon 1982a, 130, fig. 8; Dixon 1982b, 22, 25-7, no. 3; Dixon 1984, 220; *pers. comm.* Dr. T.N. Dixon.

**Oakenhead**  See **Easter Oakenhead**

A57 **Oban, log-coffin**
NM *c.* 860 307
NM83SE 6
Strathclyde Region - Argyll and Bute District
Argyll - Kilmore and Kilbride ph.

In 1878 what 'appeared...to be an old canoe' was discovered within what was probably a mortuary structure of unusual form beneath a 'mound, or rising of the land' at Dalrigh in the N part of Oban and in an area of peat-bog which was possibly a former loch. The discovery was reported at some length by Mapleton and subsequently re-assessed as a log-coffin by Abercromby, who reproduces a photograph in his account. Although there is no direct evidence for its date, it may represent a timber prototype for the type of grooved stone cist commonly found in Argyll.

The structure of 'slabs and stakes' was found 'in the peat itself'. There was 9' (2.7m) depth of peat below its bottom and 2' (0.6m) above that level; above the peat there was a deposit of 'made' ground measuring 4' (1.2m) in height and 40' (12.2m) in radius around the structure, which measured 7'6" (2.3m) in length over all, 6' (1.8m) in length internally, 2' (0.6m) in 'width' and 2'3" (0.7m) in 'depth'.

The coffin itself was 'hewn out of a solid block of oak' and aligned N-S. It measured 5'9½" (1.8m) in length by 1'7" (0.5m) in 'depth' and 2' (0.6m) in 'width'. The S end had broken away but was distinguished by grooves cut in each side to hold a board or slab. The N end was closed in a similar fashion but without a clearly-defined groove, and a possible adaptation from a previous function was indicated by the survival of part of a projecting 'keel'. Each end-slab was retained by driven stakes and further stakes along each side served to retain longitudinal logs with moss-packed interstices. Additional logs were laid parallel to the sides and obliquely at the corners while the covering comprised 'branches of birch, and probably hazel, well stuffed with moss'. Some of the timbers bore toolmarks.

Hazelnuts, split shells and charcoal were found in considerable numbers both inside and outside the coffin. Within it the soil was 'very unctuous' but contained 'no signs of bone, burnt or unburnt...and no implements of any sort'. Several pieces of birch bark (one of them probably pierced for sewing) were also found in it. One of these objects is in store at the Royal Museum of Scotland under accession number NMS EQ 192, and appears to comprise two overlapping sheets of bark which have been pierced by seven holes and bear traces of an unidentified binding material.

Mapleton 1879; Abercromby 1905, 181-2; RCAHMS 1975, 60, no. 98 (5); *pers. comm.* Mr. T. Cowie.

### A58 Perth, Saint John Street, boat
NO *c.* 119 235
NO12SW 215
Tayside Region - Perth and Kinross District
Perthshire - Perth ph.

About 1829 the 'remains of a boat' were discovered at a depth of about 10' (3.1m) during the excavation of foundations in Saint John Street which runs parallel to the present bank of the River Tay (and at a distance of 130m from it) between NO 1198 2363 and NO 1197 2346. Although the vessel was not a logboat, its unusual nature makes it worthy of re-publication.

The 'planks and bindings' of the boat were of oak while the former were both 'rabetted upon one another' and 'fastened with copper rivets'. A 'well-formed' rope of heather was attached to the boat, and beneath the vessel there were the remains of willows and aquatic shrubs.

*The Gentleman's Magazine*, (1829), ii, 267, reprinted in Gomme (ed.) 1886, i, 54.

### A59 Piltanton Burn, timber
NX *c.* 138 573
NX15NW 10
Dumfries and Galloway Region - Wigtown District
Wigtownshire - Inch and Old Luce parishes

In store at Stranraer Museum under accession number WIWMS 1956-34 there is an object which was thought to be an unfinished logboat but is most probably a length of the naturally-split trunk of a tree with part of a subsidiary branch.

It was discovered in 1945 during drainage operations along the Piltanton Burn. The archaeological recording officers of the Ordnance Survey located the discovery about 0.5km SSW of Ballancollantie Bridge and in an area at the edge of the marine sand at an altitude of about 5m OD. It was not recorded in detail at the time of discovery, but a later account classified it as an unfinished logboat measuring 9'8" (3m) in length, 2'3" (0.7m) in beam, and 1'8" (0.5m) in 'depth'. Toolmarks and areas of charred wood were noted and the timber was identified as 'oak'.

One side of the object is flattened, and, if this is placed uppermost, it measures 2.75m in length over all (of which the remaining piece of branch accounts for 0.6m), and is 0.5m broad by 0.3m deep.

Examination of the object reveals no evidence of systematic human workmanship; the isolated marks of an axe or similar tool that are visible on the flattened surface and the external charring bear no resemblance to any mode of logboat manufacture and the exterior and edges have apparently been rounded and smoothed by wear in the ground. The irregular but slightly curved hollow that is set into the flattened side measures about 1m in length, 0.1m in breadth and 0.1m in depth, and probably results from natural heart-rot. *December 1987*

McCaig 1954.

### A60 Plockton, keg
NG *c.* 80 33
NG83SW 14
Highland Region - Ross and Cromarty District
Ross and Cromarty - Lochalsh ph.

In 1888 peat-digging at a depth of 4' (1.2m) in the An Cnatharan moss revealed a 'barrel or keg hollowed out of a birch-tree trunk'; a 'cover and bottom' were 'let in'. The keg (which is stored in the Royal Museum of Scotland under accession number NMS SHC 5) contained a 'mass' of bog butter, in which cattle hairs have been identified. It measures 1'6" (0.46m) in height by 11" (0.28m) in diameter, weighs 371 lb (16.8 kg), and has a capacity of about 2 litres.

The peat-moss where the discovery was made cannot be located, but the village of Plockton is situated at sea level on the S shore of Loch Carron.

Macadam 1889, 434; Ritchie 1941, 13, 16; Earwood 1993a, 111, 285.

### A61 Polloch River, Loch Shiel, dish
NM 783 689
NM76NE 8
Highland Region - Lochaber District
Argyll - Ardnamurchan ph.

This dish was found in July 1991 in mud at the edge of the River Polloch, and at a depth of about 0.3m; it has also been published as from Loch Doilet. It is unusual in form, being roughly rectangular around the inside of the rim while the flat bottom is more nearly circular on plan and two prominent opposing flat handles project from the rim. The overall measurements are 0.3m by 0.297m, while the handles and the base measure up to 32mm and about 12mm in thickness respectively.

The dish is of alder and apparently remains unfinished. The base is roughly-worked and displays various cutmarks while the sides have numerous indistinct vertical toolmarks of a narrow and flat or slightly curved tool. Charring may be identified both within the interior and (less obviously) externally.

The vessel is currently in conservation at the Royal Museum of Scotland (under NMS daybook number DB 1991/33). It has been radiocarbon-dated to 910 ± 80 ad (OxA-3539), which determination may be calibrated to about 995 cal AD.

Hedges, Housley, Bronk-Ramsey and van Klinken 1993, 156; Earwood 1993a, 285; Earwood 1993b, 359-60, 361; *pers. comm.* Mr. T. Cowie.

## A62  Ravenstone Moss, paddles
NX *c.* 40 42
NX44SW  12
Dumfries and Galloway Region - Wigtown District
Wigtownshire - Glasserton ph.

Before 1866, five, or possibly six, paddles were discovered (probably during drainage operations) in Ravenstone Moss which is situated in drumlin country at an altitude of about 60m OD. They were close to a 'mass of timbers' which was identified as a probable crannog.

The published descriptions of these paddles are contradictory. Munro remarks that they were all of similar size, measuring 3' (0.9m) in length, but he variously cites their blades as measuring 1'2" (0.4m) and 10" (0.3m) in breadth, and 1" (25mm) and ½" (13mm) in thickness.

One of the paddles was donated to the Museum of the Society of Antiquaries of Scotland, together with the finds from Dowalton Loch (nos. 29-33), and is in store at the Royal Museum of Scotland under accession number NMS HU 14. It was recorded as measuring 2'4" (0.7m) in length, of which the handle accounts for 7" (175mm); the greatest breadth is 10" (255mm) and the greatest thickness 1" (25mm).

On examination, the most obvious feature of this surviving paddle (fig. 41) is its eccentric shape. An alternative explanation as a rudder or (less probably) a leeboard cannot be ruled out. It has apparently been made from a flat board of fine-grained and knot-free wood with rounded edges, and has become slightly warped and possibly charred in places. No distinction can be made between the ventral and dorsal surfaces.

As conserved, it measures 0.89m in length. The handle accounts for 160mm of this and measures 37mm in width by 155mm in thickness; it is squared with rounded corners in section and the end has been broken off and ground down. On one side, a pronounced reverse-curved shoulder leads into the blade; on the other, a section has been split away, giving the appearance of a slight chamfer along the unillustrated side. The surviving portion of the blade measures up to 0.25m in breadth and 15mm in thickness. *January 1988*

*PSAS*, vi (1864-6), 110; Stuart 1866a, 122-3, 150; Munro 1885, 82-4, 92, 97, 122.

## A63  River Arnol, Stornoway, Lewis, dish
NB 3007 4519
NB34NW  7
Western Isles Area
Ross and Cromarty - Barvas ph.

This carved dish of alder was found in 1986 during peat-cutting near the River Arnol. It lay inverted at a depth of 0.6m; the exterior surfaces of the dish had become weathered. No contents were recorded but what was apparently straw was noticed in and around the dish.

The dish measures 0.57m in length by up to 0.277m transversely and is elliptical on plan with slightly protruding ends and a flat base. It is of fine construction and has been carved from a split half-trunk. Rippled toolmarks within the interior suggest the use of a small gouge, but cut-marks made with a knife probably result from use rather than manufacture.

The dish is held in the Museum nan Eilean, Stornoway, and has been radiocarbon-dated to 850 ± 80 ad (OxA-3011); this determination may be calibrated to between about 910 and 977 cal AD.

Hedges, Housley, Bronk and van Klinken 1992, 144-5; Earwood 1993a, 288; Earwood 1993b, 358-9, 361.

## A64  Rough Castle, mill-paddle (possible)
NS *c.* 843 798
NS87NW  6
Central Region - Falkirk District
Stirlingshire - Falkirk ph.

The well-preserved Roman fort of Rough Castle has formed a major element of the Antonine Wall defensive system. It probably housed the greater part of the cohors VI Nerviorum for two periods within the years 142 and 164 AD approximately, and is situated on a prominent N-facing escarpment in an area where industrial activity has greatly modified the landscape. The nearest navigable water was probably in the vicinity of the River Carron which, at its nearest point, is 1.5km to the N of the fort and at an altitude about 50m lower.

The later campaigns of excavation on the fort have recently been published. The existence of waterlogged deposits or conditions suitable for the preservation of organic remains is not discussed in detail and the provenance of many of the objects is not specified, although the majority are said to be unstratified.

Among the finds there is a 'paddle-like object' of oak which has apparently suffered from splitting. As it survives, it measures 0.34m in length over all, of which the blade (of elongated oval form) accounts for 0.23m, and the rectangular-sectioned stock or handle forms the remainder. The blade is incomplete, but appears to have been roughly oval; the asymmetrical shoulders taper into the shaft which measures 70mm by 20mm in cross-section and is set at right-angles to the blade; it terminates in a neat cut which indicates that the object is preserved intact.

Although the form of this object strongly suggests an identification as a paddle, its small size casts doubt upon its possible use for propulsion; it may have been a mill-paddle.

MacIvor, Thomas and Breeze 1980, 271, no. 296; Hanson and Maxwell 1983, *passim*.


A65 **Rubh' an Dunain, Skye, oar or paddle (possible)**
NG 399 162
NG31NE 5
Highland Region - Skye and Lochalsh District
Inverness-shire - Bracadale ph.

In 1932 Scott excavated a shallow cave or rock-shelter set into the W end of Creag a' Chapuill ridge. The cave is the farthest E of the prehistoric monuments on the Rubh' an Dunain peninsula, and has entered the archaeological literature under this name. It has also been known as Rudh' an Dunain, and is situated at an altitude of about 31m OD, 200m inland from the N end of Soay Sound and about 5 km SSW of the nearest public road at Culnamean, Minginish.

The cave was found to measure about 9' (2.7m) in depth by 15' (4.6m) transversely and was exposed to the prevailing wind and rain. The remains were interpreted as evidence for Beaker period stone-knapping, Iron Age iron-smelting and transitory occupation in the Early Historic period and in modern times.

Among the finds there was a wooden blade-like object which was found on its edge about 6" (150mm) from the Iron Age furnace and apparently in the same layer. One corner and part of the surface were reported to have suffered from rot, but the remains measured 1'6" (0.46m) in length by 5½" (140mm) in breadth and ¾" (19mm) in thickness. In the ambiguous and distorted stratigraphy of the cave, it is uncertain whether this object was connected with iron-smelting activities, as Scott suggests. The published photograph suggests an identification as an oar-blade similar to that from Lochlea (no. A51) but its date is uncertain.

The artifact (fig. 42) is in store at the Royal Museum of Scotland under accession number NMS HM 301 and is identified in the accessions register as a 'spatulate wooden object'. It measures 0.45m in length by up to 140mm transversely across the blade; the blade and the diagonally-cut shoulders account for 70% of the former figure. The thickness is about 15mm throughout and there is no differentiation between the dorsal and ventral surfaces. The timber is brown in colour with burnt patches of charcoal-like texture flaking away in places, and is heavily-grained in the plane of the longitudinal axis; slight splitting has occurred. The sides, which were possibly chamfered, and the end of the blade have been worn down and rounded, giving the impression of considerable antiquity. The butt end was probably cut square across but has become irregularly worn. It is now assymetrical in form, possibly as a result of wear; on the assumption that it was formerly symmetrical, it would have measured about 160mm across.

In the absence of any evident parallel, the function of the object remains uncertain. The breadth of the blade and the shortness of the handle would, however, seem to militate against its use for maritime propulsion. *January 1988*

Scott 1934, 215, 220, 221, fig. 8; Rees 1979, i, 320-1; RMS accessions register: typescript continuation catalogue.


**Rudh' an Dunain** See **Rubh' an Dunain**


A66 **Staura Cottage, Shetland, paddle**
HU 4666 3961
HU43NE 5
Shetland Islands Area
Zetland - Lerwick ph.

In 1972 a burnt mound was excavated 70m WSW of Staura Cottage on the Ness of Sound peninsula. The discovery of pottery and iron slag indicates use in the pre-broch Iron Age. Among the finds there was 'part of a wooden shovel or paddle'; whether this was for domestic, culinary or navigational use is unclear.

*DES*, (1972), 38; Hedges 1975, 70, 75-6.


A67 **Stornoway, Lewis, trough**
NB43SW 29
Western Isles Area
Ross and Cromarty - Stornoway ph.

This carved trough was found in 1932 during peat-cutting in the Stornoway area; it lay on the old ground surface below about 3m depth of peat and the precise findspot cannot be located. It is held in Glasgow Art Gallery and Museum under accession number GAGM ARCH/NN/5, and has been radiocarbon-dated to 420 ± 80 bc (OxA-3012), which determination may be calibrated to about 406 cal BC. No contents are recorded and the timber has apparently not been identified.

The trough is similar in form to that from Loch Eport (no. A35), being of sub-rectangular form with a flat bottom, slightly thickened outwards-sloping ends, slightly rounded corners and a knob-like handle at each end. It measures 0.525m in length over all by up to 0.305m transversely. No toolmarks can be identified but the trough was of fairly crude construction.

Hedges, Housley, Bronk and van Klinken 1992, 144-5; Earwood 1993a, 50-1, 288; Earwood 1993b, 357, 358, 361.

### A68 Talisker Moor, Skye, bowl
NG 3236 3252
NG33SW 6
Highland Region - Skye and Lochalsh District
Inverness-shire - Bracadale ph.

This lugged or handled bowl of alder was found at a depth of 3' (0.9m) during peat-cutting on Talisker Moor. The vessel was found incomplete (the bottom having been penetrated, apparently by a square object) and the wood dried out considerably before conservation; the two warped major fragments and several smaller pieces are stored at the Royal Museum of Scotland (under accession number NMS IP 4). Crone notes a broad similarity to the bowl from Loch a' Ghlinne Bhig (no. A32) and suggests that they may be products of the same workshop.

The vessel is of carinated form with a sharply-inturned shoulder and an everted, facetted rim; it has two opposed prominent perforated lugs at the broadest point, and has apparently been round-bottomed. It varies between 200mm (at the rim) and 230mm (at the shoulder) in external diameter, and has apparently measured about 105mm in internal depth, suggesting a capacity of approximately 0.35 litres. The upper portion of the exterior is decorated with six ornamented grooves, each of them measuring about 6.5mm in breadth and 0.2mm in depth; beneath these there is a series of short vertical marks, each of which measures beneath 28 and 30mm in length, beneath 10 and 15mm in breadth, and up to 0.9mm in depth.

The vessel has evidently been worked by carving (rather than turning) from a length of alder (either stem or branch) measuring a little over 300mm in thickness; the billet from which the bowl was worked was placed eccentrically within this, the midpoint being located about 25mm below the rim and adjacent to one of the lugs. The groove-ornament has apparently been worked with a sharp-edge knife and the vertical grooves by a series of oblique cuts with a concave gouge.

This vessel has been dated by radiocarbon to 120 ± 80 ad (OxA-3542); this determination may be calibrated to between about 154 and 212 cal AD. The apparent similarity to some pottery forms of the later Pre-Roman Iron Age may not be coincidental. *November 1994*

Barber 1982; Hedges, Housley, Bronk-Ramsey and van Klinken 1993, 157; Earwood 1993a, 67, 289; Crone 1993b, 272-4.

### A69 Tentsmuir, paddle
NO c. 48 28
NO42NE 59
Fife Region - North East Fife District
Fife - Ferry-Port-on-Craig ph.

In September 1977 a paddle was found below high water mark on the sandy beach at Tentsmuir. It is undergoing conservation at Dundee Museum and Art Gallery under accession number DMAG 77-1050; several breaks and scratches have been infilled but only the dorsal surface could be examined on account of the fragility of the object. The timber has a fine straight grain and the shaft gives the impression of having been smoothed and polished through wear, but the object is roughly-worked and has become warped through drying. The visible scratch-marks are probably of recent date.

The paddle (fig. 36) measures 1.2m in length over all, the blade measuring 0.53m and the shaft and handle together 0.67m respectively. The blade is roughly rectangular and has a slight 'peak' roughly along the centreline; it is slightly curved in both the lateral and longitudinal planes. The extreme end of the blade is tapered on the dorsal surface and rounded ventrally. An oval hole measuring 25mm by 18mm pierces the upper central portion of the blade at a point where it is 18mm thick; it has been suggested that this may have served to secure the paddle against the side of the boat when not in use.

The shoulders of the blade are unequally concave and vary between 20mm and 40mm in thickness to form the transition into the shaft which measures about 40mm in diameter but varies considerably in section, being rounded near the blade, oval centrally and roughly square near the handle. At the end of the shaft there is a broadened handle which measures about 70mm across; this is rounded at the top and straight down the sides.

It has been radiocarbon-dated to 1420 ± 60 ad (GU-1076); this determination may be calibrated to about 1411 cal AD. *August 1987*

McGrail 1987a, 206.

### A70 Torr Righ Mor, Arran, trough
NR c. 88 31
NR83SE 12
Strathclyde Region - Cunninghame District
Buteshire - Kilmory ph.

About 1901 a 'canoe-like wooden vessel' was found during peat-digging in an area of moorland at an altitude of about 105m OD; it had probably been about 11'6" (3.5m) below the upper surface of the peat. The wood was well-preserved but serious splitting had occurred and the handle that formed one end of the object broke off and was lost during recovery operations. What are probably the remains of this vessel survive in Brodick Castle.

After recovery the object was identified as a trough which had been 'roughly hewn' from a single piece of timber; numerous toolmarks were noted. It was found to measure 4'5" (1.4m) in length by 1'8" (0.5m) transversely externally; the cavity measured up to 3'6" (1.1m) in length by 1'5" (0.4m) transversely and was apparently of rounded form in all three planes. The depth of the cavity was not recorded, but may be estimated (on the evidence of the published illustration) at about 0.2m, yielding an estimate of about 80 litres for its volume.

Although Murray hesitantly compared this discovery with the smaller logboats of Ireland and with the log-coffins of England and Scotland, it was more probably a storage vessel for bog butter.

Murray 1902; Balfour 1910, 271-2; Earwood 1993a, 102, 287.


## A71 Williamston, log-coffin
NJ 644 308
NJ63SW 9
Grampian Region - Gordon District
Aberdeenshire - Culsalmond ph.

In May 1812 a 'wooden coffin of uncommon size' was found during ploughing in an area known as Liav-park on the farm of Mill of Williamston and at a point where a cairn had been removed about thirty years before.

The coffin had been 'formed from the trunk of a huge oak, divided into three parts of unequal length, each of which had been split through the middle with wedges and stone axes, or perhaps been separated some red hot instrument of stone, as the inside of the different pieces had somewhat the appearance of having been charred'. The length of the construction was not noted but it measured about 10" (255mm) or 11" (279mm) in depth internally and was orientated E-W. It comprised six sections (two sides, two ends, a bottom and a lid) and contained an 'urn'. The lid had been splintered by earlier ploughing and the sides were found to project to a depth of 1' 1½" (0.34m) below the bottom which was retained by roughly-cut grooves. Burnt clay and ashes found beneath the structure were interpreted as the remains of a funeral pyre.

The Ordnance Survey have located the discovery at NJ 6449 3082, at a point about 170m SW of Mill Croft and at an altitude of about 130m OD in drumlin country, but study of the local topography suggests that it was more probably on the prominent Law Hillock, about 100m to the E of the recorded location. Although attributed by Wilson to his 'Iron Period', it was more probably of Bronze Age date.

*NSA*, xii (Aberdeen), 732-3; Wilson 1851, 462-3.

# IV  Synthesis and Analysis

## IV.1. Timber Supply

It is inherent in the manufacture of a boat by reduction that the form (and hence performance and functional utility) of the completed vessel will be tightly constrained by the shape and qualities (chiefly weight) of the parent log. McGrail (1978, i, 117-20; 1987a, 60, table 6.1) has cited the following types of timber as being used in European logboat construction: oak (*Quercus* sp.), alder (*Alnus* sp.), lime (*Tilia* sp.), pine (*Pinus* sp.), spruce (*Picea* sp.), elm (*Ulmus* sp.), chestnut (*Castanea sativa*), beech (*Fagus* sp.), ash (*Fraxinus* sp.) and aspen (*Populus* sp.). In general terms, they satisfy the requirements of having a long straight trunk grain allied to strength and durability, resistance to fungal decay and at least reasonable lightness. Generally speaking, the timber was probably worked unseasoned ('green') to retain a high moisture content (25% or higher), individual trees being selected for use on the basis of their size and form, freedom from knots or branches, and proximity to water or a suitable working-site.

Typically across north-west Europe, oak (McGrail 1987a, 26, 27-8, 31, 37, 40, 48, 64) appears to have been the closest to the ideal for logboat manufacture, being slow-growing and hence relatively fine-grained and easy to work. Under British conditions an open-grown unpollarded oak may achieve a girth of 6m in something over 200 years, while forest-grown tree will grow more slowly to a greater height. An oak of uncertain origin recovered from the English fenland at Adelaide Bridge and dated by radiocarbon to the mid-third millennium had a bole of length 20.4m without branches, while mature modern oaks typically have about 75mm thickness of sapwood and 25mm of bark.

Oak is quite spectacularly durable, surviving readily in natural bogs and being recovered from construction works and archaeological excavations significantly more frequently than any other species. Only its relatively high density counts against its use; with a specific gravity typically around 1, an oak log usually floats very low in the water or just below the surface, so that the finished vessel has a lower payload and is less easily portable than one made of a lighter timber. The tendency of large oak trunks to heart-rot is not in itself a problem although it may necessitate the fitting of a separate transom or (less frequently) a composite bow. It is probably too hard to have been worked with stone tools, and the experimental use of fire-working has proved ineffectual (Goodburn *pers. comm.*).

The study of reports of British logboat discoveries indicates an overwhelming predominance of examples worked from oak, although the evidence for the identification or the identity of the specialist responsible are never stated, and one may suspect that the apparent frequency of oak among the surviving remains may owe less to its frequency of use in antiquity than to the relative resistance of such hardwoods to rot in the acid conditions typical of the peat and fen deposits from which most British discoveries are derived.

This view is supported by the available evidence from Finland (NMM 1985, plates 15-17) and Denmark (Rieck and Crumlin-Pedersen 1988, 28, 41), both of which countries share some geographical similarity to parts of southern Scotland and eastern England. The ethnographic evidence for the construction of expanded logboats from Finnish aspen and the wide range of timbers (including pine, lime, ash, elm and fir) that underwater excavation has shown to have been used in Denmark, indicate the need for a re-consideration of the traditional simplistic British attribution. Indeed, Andersen (1986, 90) argues for the preferential use of lime, where available, on account of its lightness, resistance to splitting and ease of working. More recently and in a different geographic context, Earwood (*pers. comm.*) has noted the frequency of ethnographic logboat construction in willow, poplar and lime around the Baltic, on account of the resistance of these timbers to splitting and also to their light weight, which must imply a greater carrying capacity. Although the possible expansion of a hardwood (specifically oak) logboat has been experimentally demonstrated (Gifford 1993), there is no evidence for this having been achieved in Scotland. In any event, the construction and compara-

| Mowat | | Edwards | |
|---|---|---|---|
| 1 | Acharacle | | |
| 2 | Arnmannoch | 80 | Loch Rutton |
| 3 | Auchlishie | | |
| 4 | Barhapple Loch 1 | 2 | Loch Barhapple |
| 5 | Barhapple Loch 2 | | |
| 6 | Barnkirk | 3 | Moss of Barnkirk |
| 7 | Barry Links | | |
| 8 | Black Loch | 81 | Loch Sanquhar |
| 9 | Bowling 1 | 32 | Clyde 21 |
| 10 | Bowling 2 | 31 | Clyde 20 |
| 11 | Buston 1 | 5 | Loch Buston |
| 12 | Buston 2 | | |
| 13 | Buston 3 | | |
| 14 | Cambuskenneth | 7 | Cambuskenneth |
| 15 | Carlingwark Loch | 9 | Carlingwark Loch |
| 16 | Carn an Roin | | |
| 17 | Carse Loch | 56 | Friar's Carse |
| 18 | Castle Semple Loch | 85 | Loch Winnoch |
| 19 | Castlemilk | 11 | Castlemilk |
| 20 | Catherinefield | 67 | Locharbriggs |
| 21 | Closeburn | 71 | Closeburn *and* |
| | | 86 | (Royal Museum of Scotland) |
| 22 | Clune Hill, Lochore | | |
| 23 | Craigsglen | | |
| 24 | Croft-na-Caber | | |
| 25 | Dalmarnock | 45 | Dulmarnoch Farm |
| 26 | Dalmuir | 40 | Clyde 29 |
| 27 | Dernaglar Loch | | |
| 28 | Dingwall | 46 | Dingwall |
| 29 | Dowalton Loch 1 | 49 | Loch Dowalton 1 |
| 30 | Dowalton Loch 2 | 50 | Loch Dowalton 2 |
| 31 | Dowalton Loch 3 | 51 | Loch Dowalton 3 |
| 32 | Dowalton Loch 4 | 52 | Loch Dowalton 4 |
| 33 | Dowalton Loch 5 | | |
| 34 | Drumduan | 53 | Drumduan |
| 35 | Dumbuck | 39 | Clyde 28 |

| Mowat | | Edwards | |
|---|---|---|---|
| 36 | Eadarloch | 54 | Loch Treig |
| 37 | Errol 1 | 83 | Tay Errol |
| 38 | Errol 2 | 83 | Tay Errol |
| 39 | Erskine 1 | 21 | Clyde 10 |
| 40 | Erskine 2 | | |
| 41 | Erskine 3 | | |
| 42 | Erskine 4 | | |
| 43 | Erskine 5 | | |
| 44 | Erskine 6 | 41 | Clyde 30 |
| 45 | Falkirk | | |
| 46 | Finlaystone | 34 | Clyde 23 |
| 47 | Flanders Moss | | |
| 48 | Forfar 1 | | |
| 49 | Forfar 2 | | |
| 50 | Friarton | 57 | Friarton |
| 51 | Garmouth | | |
| 52 | Gartcosh House | | |
| 53 | Glasgow, Clydehaugh 1 | 24 | Clyde 13 |
| 54 | Glasgow, Clydehaugh 2 | 25 | Clyde 14 |
| 55 | Glasgow, Clydehaugh 3 | 26 | Clyde 15 |
| 56 | Glasgow, Clydehaugh 4 | 27 | Clyde 16 |
| 57 | Glasgow, Clydehaugh 5 | 28 | Clyde 17 |
| 58 | Glasgow, Drygate Street | 15 | Clyde 4 |
| 59 | Glasgow, Hutchesontown Bridge | 36 | Clyde 25 *and* |
| | | 37 | Clyde 26 |
| 60 | Glasgow, London Road | 16 | Clyde 5 |
| 61 | Glasgow, Old St Enoch's Church | 12 | Clyde 1 |
| 62 | Glasgow, Point House | 23 | Clyde 12 |
| 63 | Glasgow, Rutherglen Bridge | 35 | Clyde 24 |
| 64 | Glasgow, Springfield 1 | 17 | Clyde 6 *and* |
| | | 87 | (Royal Museum of Scotland) |
| 65 | Glasgow, Springfield 2 | 18 | Clyde 7 *and* |
| | | 42 | Additional Clyde 2 |
| 66 | Glasgow, Springfield 3 | 19 | Clyde 8 |
| 67 | Glasgow, Springfield 4 | 20 | Clyde 9 |
| 68 | Glasgow, Springfield 5 | 22 | Clyde 11 |

*Table 1. Concordance table with Edwards numbers.*

| Mowat | | Edwards | |
|---|---|---|---|
| 69 | Glasgow, Stobcross | 38 | Clyde 27 |
| 70 | Glasgow, Stockwell | 14 | Clyde 3 |
| 71 | Glasgow, Tontine | 13 | Clyde 2 |
| 72 | Glasgow, Yoker 1 | 29 | Clyde 18 |
| 73 | Glasgow, Yoker 2 | 30 | Clyde 19 |
| 74 | Gordon Castle | | |
| 75 | Kilbirnie Loch 1 | 59 | Loch Kilbirnie 1 |
| 76 | Kilbirnie Loch 2 | 60 | Loch Kilbirnie 2 |
| 77 | Kilbirnie Loch 3 | 61 | Loch Kilbirnie 3 |
| 78 | Kilbirnie Loch 4 | | |
| 79 | Kilblain 1 | 68 | Lochar Moss |
| 80 | Kilblain 2 | 68 | Lochar Moss |
| 81 | Kinross | | |
| 82 | Kirkmahoe | | |
| 83 | Knaven | 62 | Kinaven |
| 84 | Larg | 4 | Bents |
| 85 | Lea Shun | 82 | Stronsay |
| 86 | Lendrick Muir | | |
| 87 | Lindores 1 | | |
| 88 | Lindores 2 | | |
| 89 | Linlithgow, Sheriff Court-house | | |
| 90 | Littlehill | 6 | Bishopbriggs |
| 91 | Loch Ard | | |
| 92 | Loch Arthur 1 | 70 | Loch Lotus |
| 93 | Loch Arthur 2 | | |
| 94 | Loch Chaluim Chille 1 | 43 | St Columba's Loch 1 |
| 95 | Loch Chaluim Chille 2 | 44 | St Columba's Loch |
| 96 | Loch Doon 1 | 47 | Loch Doon 1 |
| 97 | Loch Doon 2 | | |
| 98 | Loch Doon 3 | 48 | Loch Doon 2 |
| 99 | Loch Doon 4 | | |
| 100 | Loch Doon 5 | | |
| 101 | Loch Doon 6 | | |
| 102 | Loch Glashan 1 | 58 | Loch Glashan |
| 103 | Loch Glashan 2 | | |
| 104 | Loch Kinellan | 63 | Loch Kinellas |

| Mowat | | Edwards | |
|---|---|---|---|
| 105 | Loch Kinord 1 | | |
| 106 | Loch Kinord 2 | 10 | Loch Canmor |
| 107 | Loch Kinord 3 | | |
| 108 | Loch Kinord 4 | | |
| 109 | Loch Laggan 1 | 65 | Loch Laggan 1 |
| 110 | Loch Laggan 2 | 65 | Loch Laggan 2 |
| 111 | Loch Laggan 3 | | |
| 112 | Loch Laggan 4 | | |
| 113 | Loch Laggan 5 | 65 | Loch Laggan 3 |
| 114 | Loch Laggan 6 | 65 | Loch Laggan 4 |
| 115 | Loch Laggan 7 | 65 | Loch Laggan 4 |
| 116 | Loch Leven | | |
| 117 | Loch nam Miol | 78 | Loch Na Mial |
| 118 | Loch of Kinnordy | 64 | Loch Kinnordy |
| 119 | Loch of Leys 1 | 1 | Loch of Banchory |
| 120 | Loch of Leys 2 | 1 | Loch of Banchory |
| 121 | Loch of the Clans | 8 | Croy |
| 122 | Loch Urr | 84 | Loch Urr |
| 123 | Lochar Moss | 68 | Lochar Moss |
| 124 | Lochlea 1 | 66 | Loch Lee |
| 125 | Lochlea 2 | 66 | Loch Lee |
| 126 | Lochlea 3 | 66 | Loch Lee |
| 127 | Lochlea 4 | | |
| 128 | Lochlea 5 | 66 | Loch Lee |
| 129 | Lochlundie Moss | | |
| 130 | Lochmaben, Castle Loch 1 | 72 | Loch Maben 2 |
| 131 | Lochmaben, Castle Loch 2 | 74 | Loch Maben 1 |
| 132 | Lochmaben, Kirk Loch 1 | 73 | Loch Maben 3 |
| 133 | Lochmaben, Kirk Loch 2 | | |
| 134 | Lochspouts | 69 | Lochspouts |
| 135 | Mabie | | |
| 136 | Milton Island | 33 | Clyde 22 |
| 137 | Milton Loch | | |
| 138 | Monkshill | | |

| Mowat | Edwards |
|---|---|
| 139 Morton | 76 Merton Mere *and* 77 Morton |
| 140 'Orkney' | 88 Orkney |
| 141 Parkfergus | |
| 142 Port Laing 1 | |
| 143 Port Laing 2 | |
| 144 Portbane | |
| 145 Portnellan Island | |
| 146 Redkirk Point 1 | |
| 147 Redkirk Point 2 | |
| 148 River Carron | 55 Falkirk Carse |
| 149 'River Clyde' | 42 Additional Clyde 2 |
| 150 River Forth | |
| 151 'River Tay' | |
| 152 Sleepless Inch | |
| 153 Stirling, King Street | |
| 154 White Loch | 75 Loch Inch |
| A1 Ardgour, bowl 1 | |
| A2 Ardgour, bowl 2 | |
| A3 Ardgour, bowl 3 | |
| A4 Ardgour, bowl 4 | |
| A5 Arisaig, Loch nan Eala, timbers | |
| A6 Bailemeonach, Mull, trough | |
| A7 Bankhead, mill-paddle | |
| A8 Bunloit, Glenurquhart, keg | |
| A9 Cairngall, log-coffins | |
| A10 Cairnside, mill-trough | |
| A11 Cnoc Leathann, Durness, trough | |
| A12 Craigie Mains, oar | |
| A13 Cunnister, Yell, Shetland, trough | |
| A14 Dalvaird Moss, Glenluce, bowl | |
| A15 Daviot, timber | |
| A16 Drumcoltran, timber | |
| A17 Dumglow, log-coffin | |

| Mowat | Edwards |
|---|---|
| A18 Eadarloch, trough | |
| A19 Easter Oakenhead, boat | |
| A20 Edinburgh, Castlehill, log-coffins | |
| A21 Edinburgh, Royal Museum of Scotland 1 | |
| A22 Edinburgh, Royal Museum of Scotland 2 | |
| A23 Glasgow, Bankton, boat | |
| A24 Gleann Geal, keg | |
| A25 Glenfield, Kilmarnock, ard | |
| A26 Gutcher, Yell, mill-paddle | |
| A27 Inverlochy, Fort William, timbers | |
| A28 Kilmaluag, Skye, keg | |
| A29 Kirkbog, paddle | |
| A30 Kyleakin, Skye, keg | |
| A31 Landis, timbers | |
| A32 Loch a' Ghlinne Bhig, Skye, bowl | |
| A33 Loch Coille-Bharr, paddle | |
| A34 Loch Doon, boat fragments | |
| A35 Loch Eport, North Uist, trough | |
| A36 Loch Glashan, paddle or oar (possible) | |
| A37 Loch Glashan, crannog, oar (possible) | |
| A38 Loch Glashan, crannog, oar | |
| A39 Loch Glashan, crannog, paddle (possible) | |
| A40 Loch Glashan, crannog, paddle (possible) | |
| A41 Loch Glashan, crannog, model paddle (possible) | |
| A42 Loch Glashan, crannog, trough 1 | |
| A43 Loch Glashan, crannog, trough 2 | |

| Mowat | Edwards |
|---|---|
| A44 Loch Glashan, crannog, trough 3 | |
| A45 Loch Kinord 5, boat | |
| A46 Loch Kinord, paddle | |
| A47 Loch Laggan, bowl | |
| A48 Loch Laggan 8, boat | |
| A49 Lochar Moss, shoulder yoke | |
| A50 Lochlea (crannog) 6, 'boat' | |
| A51 Lochlea, crannog, oars | |
| A52 Lochlea, crannog, 'double-paddle' | |
| A53 Lochlea, crannog, trough | |
| A54 Midtown, trough | |
| A55 Newstead, steering-oar | |
| A56 Oakbank, crannog, paddle | |
| A57 Oban, log-coffin | 79 Oban |
| A58 Perth, Saint John Street, boat | |

| Mowat | Edwards |
|---|---|
| A59 Piltanton Burn, timber | |
| A60 Plockton, keg | |
| A61 Polloch River, Loch Shiel, dish | |
| A62 Ravenstone Moss, paddles | |
| A63 River Arnol, Stornoway, Lewis, dish | |
| A64 Rough Castle, mill-paddle (possible) | |
| A65 Rubh' an Dunain, Skye, oar or paddle (possible) | |
| A66 Staura Cottage, Shetland, paddle | |
| A67 Stornoway, Lewis, trough | |
| A68 Talisker Moor, Skye, bowl | |
| A69 Tentsmuir, paddle | |
| A70 Torr Righ Mor, Arran, trough | |
| A71 Williamston, log-coffin | |

tive (quantitative) study of several experimental logboats worked from different timbers to a common specification is a long-overdue project.

The supposition that the logboat is a simple form of vessel, well-adapted to construction from any locally-available resources without requiring the import of expensive raw materials implies that a wide variety of types of timbers will be found to have been used across an area as large as Scotland, and that the distribution of each type of raw material will principally reflect local availability.

The availability of specific types of timber across Scotland for logboat construction has rarely been considered in an archaeological context. Richmond's view that building-timber (chiefly oak) for Roman military works in southern Scotland was imported from lowland Britain is discounted by Hanson (1978, 294-5, 297, 298) who considers that ample supplies of heavy timber would have been available, and notes that alder would have been a feasible substitute for oak, if of adequate dimensions.

The distribution of Scottish native woodland types has been summarised by McVean (1964a; b; c) with the overriding caveat that the present-day distributions of the individual types may (and probably do) vary from those of antiquity, by virtue of climatic change, soil depletion and human interference, even before the massive blanket effect of modern afforestation. The relatively recent dating that is suggested for Scottish logboats (section IV.7, below) for Scottish logboats serves to make human interference appear of greater significance than the longer-term geographical effects in the present context.

The post-glacial vegetational history of Scotland may be briefly summarised by way of introduction. In the period immediately following the withdrawal of the main ice sheets (about 15,000 bp), vegetation comprised only grasses, sedges and a few shrubs with mosses and lichens, possibly in some abundance, forming a regime similar to that now found on the plateau of the High Cairngorm. Over the subsequent two or three millennia (in the Allerod interstadial) there developed something approaching a 'park tundra' comprising thickets of scrubland trees and shrubs amidst wide swathes of heather heath. Repeated climatic cooling caused a return to more open conditions, until about 10,000 bp when climatic amelioration set in at such a rapid pace that a second period of tundra and heath has never been recognised, if it ever existed. Rather, there was a rapid colonisation by a sparse birch-dominant forest with some pine, which gradually thickened over the next two millennia, with pine increasing particularly in the lower ground. Alder and hazel increased in frequency around 5000 bc, the latter species probably causing closure of the forest canopy and the exclusion of heather, while elm and oak were present in some quantity in more southerly parts. The transition to Atlantic conditions was marked by a sudden acceleration of bog formation with increasing rainfall, until drier conditions resumed about 3000 bc, a period characterised by the much-discussed (and possibly anthropogenic) decline in elm. There is little good evidence for human interference at this period in Scotland (Smith 1981, 141-2) although clearance which may be due in part to human influence has been identified at 1685 ± 205 bc on Speyside (O'Sullivan 1974).

It may be stressed that the overall picture of Scottish forests throughout the post-glacial period is one that exhibits little variation, and certainly less than is seen in the southern English succession. Generally, birch appears to have been the dominant species throughout, with alder, hazel and pine following, while the species characteristic of oakwood were rarely prominent over much of the country.

In summary, Scotland may be conveniently divided into five zones which are considered from North to South:

### IV.1.1. THE NORTHERN AND WESTERN ISLANDS

Although the islands vary widely in the details of their cover, they are united by an over-riding unity of bogland, with montane vegetation on the few high peaks. Fragments of oak and hazel scrub in the island-chain between Arran and Skye, and birch scrub elsewhere, may be the last remnants of more widespread cover. There is no evidence of the former presence of woodland timber in the size and quantity required for logboat construction, and the boats found in this area have probably been worked from timber imported deliberately or as driftwood. The recognition (Calder 1950, 191-2; Clarke and Sharples 1985, 72) of spruce found in Neolithic settlements at Stanydale (Shetland) and Skara Brae (Orkney) as possibly of transatlantic origin may indicate the scale and nature of the available supply, while the interpretation of the 'temple' (house) at the former site as requiring some 700m length of dressed timber for its construction must serve as a warning against assumptions of a total unavailability of timber, even in what is now the treeless far North.

Pollen studies on these islands (Tipping 1994, 11-13) are complicated by the problem of assessing the quantity and significance of allochthonous wind-blown pollen. It is probably wrong to assume that extensive areas of closed-canopy coverage were to be found, open scrubland being probably more typical. Nevertheless, some of the abundant pine stumps on Lewis and Harris have been dated between 2900 and 1900 BC, while there is pollen evidence for small stands of oak, elm alder and, after 2450 BC, ash. On Shetland, pollen evidence suggests at least local natural cover of birch, hazel, juniper and rowan, which were possibly accompanied by oak, alder, elm and ash. In the case of Orkney, it has been suggested that the woodland of birch, hazel, willow and, possibly, alder was disturbed after about 4550 BC, allowing colonisation by a greater diversity of taxa, including oak and pine. Radiocarbon dates for wood cluster between 6800-5500 BC and 3300-1800 BC in the Western Isles and at comparable dates on Shetland.

Tipping (1994, 23-5) sees trees as only having a tenuous hold in this area throughout the early and middle Holocene. The pollen evidence suggests that the area supported open birch/hazel woodland for the most part. Anthropogenic effects have been distinguished in some localities, notably on Orkney at about 3000 BC and in Shetland where the beginning of agricultural activity has been recognised at about 2750 BC.

### IV.1.2 THE NORTHERN HIGHLANDS

Within mainland Scotland and to the North of a line drawn between Strathfarrar and Loch Carron, the prevailing vegetation is blanket bog of varying types, with intermittent birch forest and scrub, and juniper scrub in places. Exceptionally, the calcareous areas of west Sutherland and Wester Ross support only a specifically adapted sward and occasional ash woods, while the Flow country of Sutherland has probably always contained extensive unforested areas on account of its wetness; pine woodland may be found in the southern part of the zone.

Tipping (1994, 13-14) argues that present-day altitudinal limits are largely artificial, having been lowered by climatic change, grazing pressure and the development of upland blanket peat. The maximal extent of tree cover is seen as clearly declining towards the N and W, under the influence of climatic stress. It is possible that over half the altitudinal range in the NW highlands may always have been treeless.

On the basis of pollen evidence, Tipping (1994, 23-5) sees trees as only having a tenuous hold in this area throughout the early and middle Holocene. The 'flow country' and northern Scotland in general, probably supported open birch/hazel woodland for the most part. Under these conditions, small-scale changes in the landscape may have effectively inhibited regeneration, making anthropogenic effects difficult of recognition, although the beginning of cultivation has been recognised (at about 3000 BC) on the NW coast.

### IV.1.3 THE EAST-CENTRAL HIGHLANDS

The Grampian Massif to the East of Rannoch Moor is dominated by heather moorland, with the specific type varying with altitude, and montane vegetation on the peaks. Altitudinal variation is pronounced throughout this zone.

Intermittent pine forest survives throughout the zone and the ever-present peat bogs have been in part derived from the felling of such woods, which formerly extended to an altitude of about 610m OD. The forest cover is most pronounced in the valleys of the River Dee and of the rivers flowing into the Moray Firth, notably the Spey, along the banks of which it was commercially exploited, at least on a small scale, as early as the early eighteenth century (Fenton 1972, 70-1). Oak is found at least sporadically and juniper is the most common scrub throughout. Deciduous timber of adequate dimensions for logboat construction is a rarely-found commodity.

Here, as elsewhere, Tipping (1994, 13) argues that present-day altitudinal limits are largely artificial, having been lowered by climatic change, grazing pressure and the development of upland blanket peat. The present-day altitudinal limit to pine at Creag Fhialach on the W flanks of the Cairngorm massif is thought to be the only surviving natural tree-line in Scotland at 648m OD.

On the basis of pollen evidence, Tipping recognises (1994, 29-30) a woodland dominated by birch and hazel with small amounts of oak and elm along the E and NE coasts to the N of the Firth of Tay. These do not appear compatible with the open birch/hazel woods of northern Scotland, but are, by contrast, of closed-canopy form. Proportions of oak and elm decline with both altitude and latitude towards the limit of their range.

Unfortunately this region has seen less research than other parts of Scotland, radiocarbon-dated pollen analyses having been undertaken in a lowland context only in the Howe of Cromar. This area lies close to the boundary with the pinewood area and need not be typical of the region as a whole. Small-scale human interference has been identified at about 3350 BC and in subsequent episodes, while a major phase of activity between 1100 and 300 BC was accompanied by the earliest evidence for both cereal cultivation and soil erosion. By contrast, the pollen evidence from the uplands to the S of Deeside attests to small-scale activity (possibly transhumant grazing) in later prehistory.

*IV.1.4. THE WEST-CENTRAL HIGHLANDS*

The Western Highlands vary subtly from the area to their East by virtue of the relative absence of forest. Mixed grassland and dwarf shrub communities now predominate, heather having been largely eradicated by burning and sheep-grazing. Bracken is now widespread, and the intermittent forest cover is of oak and birch, which were apparently of considerable extent before the extensive clearance associated with eighteenth-century iron-working (Anderson 1967, i, 453-60).

Tipping (1994, 25-9) sees this area as being the heartland of the pine/birch ('native pine') woodlands which appear to have spread across the already-developing peat from about 300 BC. No evidence for early anthropogenic influence has been recognised in an area which presumably had a very low population and which shares the problems of interpretation noted in the (Outer) Western Isles (section IV.1.1, above). To the S and E of the Great Glen, the pinewoods were apparently more closed and a more complex history of clearance is apparent. The almost complete cover of blanket bog and rare birch copses that is typical of the area probably developed without human interference. Although pine advanced into the northern mainland about 2850-2450 BC, it apparently collapsed relatively soon afterwards, at about 2000-1800 BC, whether as a result of human interference, climatic change or fallout of Icelandic volcanic tephra.

Pine appears never to have been abundant on the W coast to the S of Loch Maree, being displaced by oak in mixed birch/hazel/oak woods in which elm was locally present. Widespread human influence has been recognised after about 2200 BC, although more localised disturbance has been recognised from as early as about 3250 BC and particularly around 2600-2500 BC.

*IV.1.5. THE CENTRAL LOWLANDS, STRATHMORE, THE BORDERS AND THE SOUTH-WEST*

This area of oak and birch woodland has been altered more than any other by deforestation, clearance and emparkment, but has apparently traditionally supported extensive oakwoods, of which semi-natural remains survive in places. Grassland and grassland-bogs with a few dwarf shrubs are found on the hills of the zone. The area was probably oak forest up to an altitude of about 300m, with zones of pine and birch above, and, as such, it apparently offered the best supply of raw material in Scotland, a supposition which is supported by the recorded distribution (section IV.2, below).

Particular interest attaches to the distribution of oak woodland, in view of its traditional status as the preferred raw material for logboat manufacture and its general under-representation in the palynological record on account of the small quantity of pollen typically produced by this species. Today, it is found sporadically as far north as Letterewe, Wester Ross, but the surviving reserves at Clais Dhearg, near Taynuilt in Lorn, at Airds Wood in Appin, and at Ariundle in Ardnamurchan are more typical of the surviving oakwoods of the Western Highlands, in which area oak was apparently the climax vegetation, and of which the present woodlands are degraded remnants. Oaks of sufficient size for boatbuilding and other timber construction must always have been particularly rare within the Highlands, where Cheape (1993, 56, 58) notes the tradition of galleys having been built at Kishorn, on the shores of Loch Awe and Loch Fyne, in Skye, on Barra and at Gasgan (on Lochshielside). In the construction of these vessels, naturally-bent oak was used for the structural timbers and elm for the strakes; the last-cited location was chosen for its proximity to the best available oakwood and the same factors was presumably important elsewhere. Within the post-medieval period, Smout (1993, 46-7) has cited documentary evidence for species-specific woodland management, with the particular objective of maintaining oak resources in Ardnamurchan, NW Argyll and SW Perthshire as a basis for charcoal-burning and tanbarking. Oak was preferred to ash, rowan and holly in these areas while elsewhere eighteenth-century accounts tell of the destruction of self-regenerating young oaks by both animals and man.

Tipping (1994, 13-14, 18, 30-2) argues that present-day altitudinal limits are largely artificial, having been lowered by climatic change, grazing pressure and the development of upland blanket peat. In the Southern Uplands, an open and diverse tree cover has been recognised up to about 600m OD, a figure which accords well with comparable contexts in Northern England and suggests that no unforested area need have existed to the S of the Forth-Clyde line at the maximum extent of woodland. Woodland was both more dense and more diverse away from the exposed W coast, with closed-canopy conditions predominating; short-lived clearance episodes which need not be anthropogenic in origin.

This area is characterised by oak/hazel/elm woodland in which birch was present but subsidiary, being possibly shaded out by the canopy cover. This forest was best developed in the great expanse of low-lying ground between the Highland Boundary and Southern Upland faults. Some variations in composition may be recognised, neither oak nor elm being thought to have colonised the base-poor soils of the Campsie Fells while the acid Silurian rocks of the upper Southern Uplands discouraged colonisation by elm. Early clearance has been recognised across the southern part of this area at about 2000-1850 BC with some later regeneration after about 800 BC.

In general terms and across the country as a whole, the distribution of logboat discoveries accords well with the recorded distribution of oakwood at the present day and in the recent past, and the size of vessel typically found with the relatively small size of trunk that is to be expected near the periphery of the distribution of the species.

The distributions of the other species known to have been used for the construction of European logboats (McGrail 1978, i, 117-20) may be briefly discussed. Apart from its widespread commercial prevalence, pine is found naturally (McVean 1964a, 146-50) as major forests in the valleys around the Grampian massif and immediately north of the Great Glen on well-drained sands and gravels up to an altitude of about 500m. Ashwood has probably (McVean 1964a, 158-9) formed the climax vegetation in the restricted limestone and calcareous areas of the western highlands and islands to the south of Sutherland. It is not found in Durness but survives in pure stands at the head of Loch Kishorn and on Sleat, while there are smaller quantities on Lismore. Alder is found (McVean 1964a, 160) in swamps on level alluvial ground and hillslope woods in areas of extensive water seepage and springs, the classic example being in the area around Urquhart Bay, Loch Ness. Aspen (McVean 1964a, 157) is also found sporadically in the forest zone. Lime, chestnut and beech are not found naturally in Scotland, and elm rarely so, while spruce is a recently-introduced timber which is characteristic of commercial woodlands.

## IV.2. Geographical Distribution and the History of Discovery

The study of archaeological distributions has frequently been criticised as a potentially misleading technique on the grounds that the picture presented is that of recorded discovery rather than original presence. Recorded examples may be located atypically for reasons of differential preservation (through soil chemistry, topography or land use), differential discovery (through variations in modern human activity) or differential recording (through the availability of archaeological resources or variations in the significance of the artefact-type as seen by successive scholars). These factors are discussed with particular relevance to maritime archaeology by Muckelroy (1978, 157-213, *passim.*) while the varying significance of logboat discoveries in the development of Scottish archaeological research and thought forms a subject for future study.

This qualification is of particular significance in the study of the logboat, a class of artefact which requires to be recognised at the time of its discovery, or at least soon afterwards, if it is to be comprehensively recorded. Logboats are typically first revealed as ill-defined and unattractive pieces of timber which bear little evidence of human workmanship and may be only partly-exposed at first. They are of less evident interest or aesthetic appeal than artefacts of such types as metalwork or even pottery, but, perversely, they tend not to survive if not immediately recognised, as a result of disintegration resulting from drying-out on removal from their long-term environment, while their relatively small size means that they are less likely to survive undamaged than such larger (and partially stone-built) structures as crannogs and island-dwellings. Although it is impossible to quantify such losses, we may assume that in any land drainage, peat-digging or similar operation, the crannogs would have had a reasonable chance of survival as 'islands' projecting above the dug ground, and any recognisable artefacts would have been removed for sale, record or both, while logboat remains would have stood the greatest chance of being cast aside unrecognised to disintegrate or be burnt along with the ancient unworked timbers frequently uncovered in such activities.

Criteria for the identification of logboat remains have been laid down by McGrail (1978, i, 19; 1987a, 56-7) but in the absence of such guidance coupled with supervision by knowledgeable informants, it is likely that the recorded examples form only an extremely small proportion of those discovered. Collateral English evidence for this assertion may be provided by the recorded discovery of only six logboats in the Somerset Levels, an extensive area which is topographically well-suited to their construction and use, which has a long tradition of archaeological field work, and over which peat extraction has been carried on such a scale that the upper layers have been largely removed (McGrail 1978, i, 178, 195-9, 238, 269-70, 302; Coles and Coles 1986, 14-16, 31, 114-17, 180).

By comparison, the major concentrations of logboat discoveries in Scotland result from repeated site-watching by one or more knowledgeable individuals within a relatively small area. The most noteworthy of these was the often-hurried recording of logboats that was carried out on discoveries made during the extensive river-works and harbour-construction that took place on the upper Clyde during the later nineteenth century (Riddell 1979, *passim.*) and which may be seen in retrospect as one of the earliest coherently-organised rescue archaeology projects. The major figure in this work was John Buchanan, a major Glasgow lawyer, banker and antiquary ('Senex' 1884, iii, 502-3) who lived from 1802 to 1878. Being, apparently, of a naturally-retiring disposition, he normally wrote anonymously, or nearly so, over the subscriptions of 'JB' or 'Aesica', and it is nearly impossible to determine whether he was responsible for the initial recording of a specific example or has given a secondary account. He certainly achieved the publication of seventeen logboats and the Bankton boat in his three major articles (Buchanan 1848; 1854a; 1884), and probably personally recorded most, if not all, of those discovered between 1848 and 1868. His last summary article reveals his mode of thought to have been advanced for its day; the question of the date of construction is considered in some detail, even if his lack of analysis of the morphological characteristics led him to attribute an excessively early date, while the geological situation, cultural background and mode of construction are considered at some length with the assistance of classical sources. This work is only paralleled outside Scotland by the successive discoveries noted by Madely, Dunlop, Dale and Leigh at Warrington on the River Mersey (McGrail 1978, 287-98; McGrail 1979).

After Buchanan's death, the work of recording discoveries in the Clyde passed briefly into the less productive hands of Murray, Bruce and Donnelly, before ceasing entirely in the early years of the present century, presumably with the result that numerous boats were destroyed without record in the later years of dock construction.

The baseline for the nineteenth-century study of the logboat in Scotland was established by Daniel Wilson (1816-92), whose place as an archaeological theoretician and synthesiser alongside Worsaae has been noted by Simpson (1963) and Daniel (1981, 48, 94, 162); his achievements as a historian and illustrator of Edinburgh architecture, as an ethnographer and in the adminstration of Canadian education were no less real.

In the first edition (1851) of his seminal work he summarises the known logboat discoveries in a chapter of 'Aboriginal Traces' which also covers carseland whales and harpoons, peat-moss wheels and (presumed Roman) trackways, as well as miscellaneous timbers and trunks. Logboats are considered of importance as 'the first colonist of the British Isles must have been able to construct some kind of boat, and have possessed sufficient knowledge of navigation to steer his course through the open sea' (1851, 29). Logboats were assumed to be vessels of early seaborne trade and of initial colonisation while the survival and recording of their remains is treated with characteristic grandiloquence: 'Time has dealt kindly with the frail fleets of the aboriginal Britons, and kept in store some curious records of them, not doubting but these would at length be inquired for' (1851, 30). Rarely can the Victorian assumption that all phases of prehistory can be illuminated by the surviving archaeological record have been so clearly stated.

At this point, his line of argument loses touch with nautical probability as he suggests that 'It is by no means to be presumed as certain that the early navigators chose the Straits of Dover as the readiest passage to the new world they were to people ... Whencesoever the first emigrants came. Providence alone could pilot their frail barks'. Certainly, providence would have a major role to play in any successful coastal voyage made in any of the examples that he cites. The River Carron ('Carron Valley') discovery of 1726 (no. 148) is stated to be 'of larger dimensions than any other discovered to the north of the Tweed', but measured only 11m in length by 1.4m in beam and 1.3m in depth, dimensions considerably smaller than the corresponding figures (about 12.6m by 1.5m by 1.2m) for the Hasholme boat which has been seen as a coastal craft operated within the extensive estuary of the River Humber (Millett and McGrail 1987, figs. 12-13 and 24-25).

Comparison with other records (with the benefit of hindsight) shows that Wilson was in touch with events in the Clyde valley and the South-west, but was unaware of discoveries in the Western Isles and in the North and East. Notable discoveries omitted from his account, although in print by 1851, include Loch Chaluim Chille 1 (no. 94), Loch of Kinnordy (no. 118), Loch of the Clans (no. 121), and the boats from Barry Links (no. 7), Lindores (nos. 87-8) and Sleepless Inch (no. 152) in the Tay estuary. Presumably, the apparently-early dating of the Kinnordy discovery on geological grounds would have seemed particularly significant to him.

The second edition of Wilson's work was partially rewritten, but in the corresponding chapter (1863, i, 41-61) there are no additional logboat discoveries recorded. The deeply-buried skull from Grangemouth is first mentioned, as is the clinker-built boat from Glasgow, Bankton (no. A23) while the flint discoveries by Boucher de Perthes in the river-terrace gravels of the Somme and elsewhere in Northern France are discussed at length, but the Glasgow, Clydehaugh discoveries of 1852 (nos. 53-7) are omitted as are those from Erskine 1 (no. 39), Glasgow, Point House (no. 62) and Port Laing (nos. 142-3). He evidently felt that the book was not intended as a complete and updated summary of Scottish archaeological knowledge, and there was no reason to revise the argument stated in the first edition.

The second of the major 19th-century archaeologists to consider the logboat an important type of artefact was John Stuart (1813-77) who is best known for his prominent role in the affairs of the Society of Antiquaries of Scotland (Clarke 1981, 127-8; Stevenson 1981a, 84; Stevenson 1981b, 148, 156-7), but who early realised the prevalence and significance of lacustrine discoveries in Scotland (particularly those made in drained areas) and combined the results of recent work at Dowalton Loch with that of Joseph Robertson at Loch of Leys to produce a synoptic article (Stuart 1866). His major interest in doing so evidently lay in the crannogs, lake-dwellings and fortified islands that are discussed at length with the assistance of parallels from the newly-discovered Swiss lake-villages and from Irish and Scottish historical accounts. The evidence for medieval occupation is rehearsed at some length and crannogs are seen as essentially temporary refuges. Logboats are noted, like discoveries of metalwork and organic remains, in terms of their major dimensions, and some are included that are not strictly lacustrine, notably those from the River Clyde. Monoxylous boats are clearly differentiated from those of skin construction (1866, 148-52) and are compared with the recorded Irish examples, while Stuart is the only authority to pay attention to the recorded discoveries of paddles.

Arguably the most eminent archaeologist to be concerned with the recording of Scottish logboats was Robert Munro (1835-1921) who, after a medical career in Ayrshire, devoted himself to the study of what would now be termed wetland archaeology (Munro 1921, *passim.*), first through a series of thoroughly competent and well-published excavations on crannogs and waterlogged sites in Ayrshire, Dumfries and Galloway, and, after 1885, through the compilation of a summary guide to the equivalent discoveries of whatever period, across Europe (Munro 1890). He also essayed an early account of Scottish prehistory as a whole (Munro 1899) in which wetland sites are set in their true perspective. His first and third contributions to archaeology closely parallel the contemporary work of Wood-Martin (1886; 1895) in Ireland, but his European work is unique.

Although his account of the Buston 1 logboat (no. 11) was comprehensive, percipient and proficient, he was not concerned with boats as such, but was primarily interested in the structures, functions and contents of crannogs; such logboats as were incidentally discovered were recorded on that basis. He was sufficiently astute to realise that logboats were not necessarily to be seen as indicators of crannogs, but could exist separately or in association with such later structures as castles. Indeed, he was quick (1882, 279-80) to doubt the traditional prehistoric attribution of logboats and to suggest a later dating, in contradistinction to the traditional view as expressed by Stuart.

In his final survey of Scottish prehistory, Munro (1889, 66-73, 359-61) avoids this question and makes no attempt at a complete summary of the evidence or at a study of their distribution. He follows Wilson in noting the discoveries of logboats from apparently-early contexts in the Carron Valley and near Falkirk, and in the peat-mosses to the south-east of Dumfries; these latter discoveries are also compared (following Mitchell) to the buried oaks of the Forest of Cree. The geologically-early discovery from Friarton is mentioned and a brief description of some of the Clyde examples concentrates on the apparent high antiquity of those from the upper river terraces. In a later chapter, the logboat is contrasted against the skin boat, but no chronological or cultural conclusion is drawn.

The work of these illustrious Victorians bequeathed to Scottish archaeology a distribution of recorded logboats which has remained essentially unchanged to this day, and is described below. However, it is instructive that the subsequent generation of scholars considered logboats as being of far less significance, and accorded them correspondingly less attention. The archaeological establishment of the inter-war years was preoccupied with other topics: Macdonald re-directed attention to Roman matters and was followed in this by Callander who investigated (1929) the nature of sea-level changes around Scotland in response to the discovery of an alleged, but disproved, logboat in the inner Forth estuary. He only considered one logboat worthy of mention, namely, the Dumbuck example, which was noted with the dock structure and the nearby crannog. He evidently considered that submerged forests, flint discoveries, shell-middens, and, on occasion, chambered cairns and later prehistoric fortifications had replaced logboats as indicators of geomorphological change.

Even Childe, for all his preoccupation with trade and population movement, displayed only a minimal interest in the waterborne mechanics of these processes. In his summary account of Scottish prehistory, 'canoes' are only mentioned (1935, 18) in an aside which casts doubt on the antiquity of the Forth valley examples, while in his Rhind lectures of 1944 (published as Childe 1946) such artefacts are ignored entirely. In his more wide-ranging account of British prehistory he mentions logboats (1940, 130, 174, 237) only to note their occasional re-use as coffins, the existence of a transom-groove on one example, and their use as a means of local transport around the Iron Age settlement at Glastonbury.

This approach was also followed by Scott (1951, 16, 30) who began his study of the Neolithic 'colonisation' of Scotland with the assertion that 'the first problem of the trader is to secure the right to trade', thus implying the improbable assumption that contemporary maritime technology presented no limitations to a process which he saw, on the basis of parallels from colonial America and Canada, as primarily waterborne. He avoided the problem of the small size of recorded British logboats by proposing the existence in prehistory of a larger, extended version for which no evidence survives. He obviously considered the recorded distribution of logboats so incomplete that it need not be considered as evidence for or against the use of specific natural routeways.

The major sources of new discoveries in the foreseeable future are likely to be sport diving and gravel-digging, with a resultant shift in the recorded distribution. Sport diving has been technically feasible in Scotland since soon after the Second World War but has grown rapidly in popularity over the last twenty years, particularly in the large inland lochs of Argyll and the Trossachs which offer clear and unpolluted non-tidal water in pleasant scenery and within reasonable travelling distance of the major population centres.

During the same period gravel-digging has become a major agent of both destruction and discovery in Scotland as in England, both on account of the considerable areas of river flood-plain excavated to provide aggregate building material, and the excavations conducted to achieve levelling and provide core material for motorway and road construction. Recent discoveries from both these sources are described in the gazetteer, and more may be expected.

The result of this changing pattern of discovery will most probably be to thicken up the recorded distribution in the inland lochs of Argyll and the Trossachs and, less noticeably, in the central belt and the sea-lochs of Argyll and the lower Clyde. It is unlikely, however, that sport divers will ever adequately explore the sea-lochs of the North-West highlands or the numerous lochs of Badenoch and the Grampian massif, to say nothing of the far north. There remains a place for at least localised archaeological survey to confirm or deny the hypothesis that most highland lochs may be expected to contain logboats in the numbers indicated by the numerous discoveries in Loch Laggan.

Indeed, some areas which now appear surprisingly bereft of logboats will probably remain so. The absence of recorded examples in the numerous lochans of the Lothian coastal plain and the eastern Borders is surprising, as is their paucity in Buchan and the north-east. The two examples recorded from Orkney are sufficient to indicate that such boats can be built in the far north and that discoveries remain to be made in Caithness and the less mountainous parts of Sutherland.

Further background work might usefully also include the detailed investigation of a small loch and its surrounding peat, employing coring, probing, excavation and other techniques of environmental and geomorphological research in the hope of revealing structures, logboats and metalwork deposits and placing them within a dated sequence of stratified deposits.

The present recorded distribution (fig. 1) can be most easily analysed on the basis of seven geographical groups, which are first tabulated and considered in their topographical and archaeological contexts before an attempt is made to analyse the constructional variations found in each:

*IV.2.1. DUMFRIES, GALLOWAY AND AYRSHIRE*

(53 discoveries)

| | | |
|---|---|---|
| 2 | Arnmannoch | pre-1901 |
| 4 | Barhapple Loch 1 | 1880 |
| 5 | Barhapple Loch 2 | 1884 |
| 6 | Barnkirk | *c.* 1814 |
| 8 | Black Loch | *c.* 1861 |
| 11 | Buston 1 | 1880-1 |
| 12 | Buston 2 | 1992 |
| 13 | Buston 3 | undated |
| 15 | Carlingwark Loch | poss. 1765 |
| 17 | Carse Loch | pre-1869 |
| 18 | Castle Semple Loch | pre-1795 |
| 20 | Catherinefield | 1973 |
| 21 | Closeburn | 1859 |
| 27 | Dernaglar Loch | 1885 |
| 29-33 | Dowalton Loch 1-5 | 1863-4 |
| 75 | Kilbirnie Loch 1 | 1868 |
| 76 | Kilbirnie Loch 2 | post-1868 |
| 77 | Kilbirnie Loch 3 | 1930 |
| 78 | Kilbirnie Loch 4 | 1952 |
| 79 | Kilblain 1 | 1736 |
| 80 | Kilblain 2 | 1772 |
| 82 | Kirkmahoe | 1919 |
| 84 | Larg | pre-1863 |
| 92 | Loch Arthur 1 | 1874 |
| 93 | Loch Arthur 2 | 1966-7 |
| 96-7 | Loch Doon 1-2 | 1823 |
| 98-101 | Loch Doon 3-6 | 1831 |
| 122 | Loch Urr | pre-1927 |
| 123 | Lochar Moss | pre-1791 |
| 124-5 | Lochlea 1-2 | *c.* 1840 |
| 126-8 | Lochlea 3-5 | 1878-9 |
| 130 | Lochmaben, Castle Loch 1 | 1909 |
| 131 | Lochmaben, Castle Loch 2 | 1949 |
| 132 | Lochmaben, Kirk Loch 1 | 1910 |
| 133 | Lochmaben, Kirk Loch 2 | 1911 |
| 134 | Lochspouts | *c.* 1875 |
| 135 | Mabie | pre-1879 |
| 137 | Milton Loch | 1953 |
| 139 | Morton | early 18th century |
| 146 | Redkirk Point 1 | 1954 |
| 147 | Redkirk Point 2 | 1956 |
| 154 | White Loch | *c.* 1870 |

The largest group differentiated is that comprising 53 discoveries from Dumfries, Galloway and Ayrshire. For the most part, these examples were recovered during moss or loch drainage during the improvement phase of the eighteenth century, in the last century, or (less frequently) in more recent years, and have been only briefly noted by inexperienced amateurs. Those discovered and recorded during crannog excavations by Lovaine, Munro and Piggott, and those revealed during industrial activity around Lochwinnoch and Kilbirnie are notable exceptions.

The distribution of discoveries within the area is a fairly uniform scatter around the altitude band between 10 and 90m OD in the cultivated clayland and drumlin country around the edge of the Galloway hill massif, although concentrations in upper and middle Nithsdale, southern Ayrshire and the Lochwinnoch area reflect the activities of Munro and his correspondents. Exceptions to this are the single boats found at Black Loch and at Mabie, and the groups at Loch Doon and at Lochlea. The absence of recorded examples in the extensive mudbanks of the deeply-embayed Solway coast is noteworthy, particularly as most of the shore must have been continuously walked by the workers at the numerous salmon-fisheries that characterise of this area.

Only two possible marine examples have so far entered the literature, and it may be surmised that future discoveries in this area will be predominantly maritime. Continuing and repeated land drainage operations will probably reveal further examples, as will sports diving in some of the inland lochs, notably Loch Ken and Clatteringshaws Loch.

*IV.2.2. Clyde Estuary and River-terraces*

(34 discoveries)

| | | |
|---|---|---|
| 9-11 | Bowling 1-2 | 1868 |
| 26 | Dalmuir | 1903 |
| 35 | Dumbuck | 1898 |
| 39 | Erskine 1 | 1854 |
| 40-2 | Erskine 2-4 | 1893 |
| 43-4 | Erskine 5-6 | 1977 |
| 46 | Finlaystone | 1878 |
| 53-7 | Glasgow, Clydehaugh 1-5 | 1852 |
| 58 | Glasgow, Drygate Street | pre-1848 |
| 59 | Glasgow, Hutchesontown Bridge | 1880 |
| 60 | Glasgow, London Road | 1825 |
| 61 | Glasgow, Old St Enoch's Church | 1780 |
| 62 | Glasgow, Point House | 1851 |
| 63 | Glasgow, Rutherglen Bridge | pre-1880 |
| 64 | Glasgow, Springfield 1 | 1847 |
| 65-6 | Glasgow, Springfield 2-3 | 1848 |
| 67-8 | Glasgow, Springfield 4-5 | 1849 |
| 69 | Glasgow, Stobcross | 1875 |
| 70 | Glasgow, Stockwell | 1824 |
| 71 | Glasgow, Tontine | 1781 |
| 72-3 | Glasgow, Yoker 1-2 | 1863 |
| 136 | Milton Island | 1868 |
| 149 | 'River Clyde' | undated |

The second largest group to be recognised is that from the Clyde estuary and its river-terraces; this comprises some 34 discoveries, the great majority of them made during dock construction and river works before the first world war. Many of these discoveries were made before the development of the modern system of archaeological recording, museums and conservation facilities and so could not be recovered or preserved, with the result that only 7 (20% of the group) now survive.

The steady conversion of the Clyde from navigational to recreational use, combined simultaneously with encroachment by waterside development, natural silting and the deliberate infilling of docks and harbour works, makes this archaeological environment one of material deposition and will probably preclude further logboat discoveries in the classic area on the west side of Glasgow. However, further discoveries can be expected through river erosion in the areas below and behind the extensive training walls that delineate the channel around Dumbarton, and possibly through the activities of sports divers in the extensive series of drowned sea-lochs (Loch Fyne, Loch Striven, the Gare Loch and the Holy Loch) that open onto the lower Clyde. Further discoveries probably yet remain to be made during building and construction works in the upper river terraces, where any boats found will be of exceptional interest on account of their probable early date.

*IV.2.3. Highland lochs and bogs (including Argyll, the Trossachs and the Western Isles)*

(27 discoveries)

| | | |
|---|---|---|
| 1 | Acharacle | 1895-1900 |
| 16 | Carn an Roin | 1972 |
| 24 | Croft-na-Caber | 1994 |
| 25 | Dalmarnock | 1975 |
| 28 | Dingwall | 1874 |
| 36 | Eadarloch | post-1938 |
| 91 | Loch Ard | 1986 |
| 94 | Loch Chaluim Chille 1 | 1763 |
| 95 | Loch Chaluim Chille 2 | 1874 |
| 102 | Loch Glashan 1 | 1960 |
| 103 | Loch Glashan 2 | 1961 |
| 104 | Loch Kinellan | 1915 |
| 105 | Loch Kinord 1 | 1858 |
| 106-8 | Loch Kinord 2-4 | 1875 |
| 109-10 | Loch Laggan 1-2 | 1934 |
| 111 | Loch Laggan 3 | 1948 |
| 112-13 | Loch Laggan 4-5 | 1949 |
| 114-15 | Loch Laggan 6-7 | 1955 |
| 117 | Loch nam Miol | *c.* 1870 |
| 141 | Parkfergus | 1790 |
| 144 | Portbane | *c.* 1977 |
| 145 | Portnellan Island | 1913 |

The recognised and recorded discoveries from this vast area of deep glaciated valleys, drained lochs and bogs, and highly-indented coastline undoubtedly represent only a minute proportion of those actually made, as this area has traditionally received the lowest level of archaeological surveillance of any in Britain, if not in Europe. The relatively large numbers of discoveries made in those lochs (notably Glashan, Kinord and Laggan) that have received a degree of systematic surveillance indicate the density of discoveries that probably remain to be made elsewhere.

The pattern of logboat discoveries parallels that of crannogs, particularly in respect of the lack of recorded examples from marine contexts. Only the boat found at Acharacle (no. 1), and possibly also those from Loch Chaluim Chille (nos. 94-5), appear to have seen maritime use. Comparison may be made within this area with the record from Argyll of a total of 72 crannogs and related structures (RCAHMS 1971, 94-5; 1975, 93-5; 1980, 119-23; 1984, 153-7; 1988, 205-8; 1992, 302-6) of which only three are in marine situations. Of the others, 24 are in major glaciated valley-lochs, 23 in small hill-lochs, and the remaining 14 in the small lochs that are characteristically interposed among the machair sands. It is significant that twenty of these crannogs were discovered during a single season of underwater survey in Loch Awe (Hardy 1973; Morrison 1985, 31-3). The development of sport diving and systematic underwater survey (using remote survey technology wherever possible) offer the best prospects for an increased rate of discovery in future years, although marine examples will probably remain under-represented.

*IV.2.4. STRATHMORE, ANGUS, THE MEARNS, AND THE NORTH-EAST*

(14 discoveries)

| | | |
|---|---|---|
| 3 | Auchlishie | 1791-1820 |
| 23 | Craigsglen | c. 1893 |
| 34 | Drumduan | 1836 |
| 48 | Forfar 1 | c. 1862 |
| 49 | Forfar 2 | 1952 |
| 51 | Garmouth | 1886 |
| 74 | Gordon Castle | c. 1886 |
| 83 | Knaven | 1850 |
| 118 | Loch of Kinnordy | 1820 |
| 119-20 | Loch of Leys 1-2 | 1850 |
| 121 | Loch of the Clans | c. 1823 |
| 129 | Lochlundie Moss | pre-1867 |
| 138 | Monkshill | 1889 |

Most of the discoveries in this widely-spaced group were made during agricultural drainage and clearance operations during the improvement period. Only one boat in the group has been discovered during the present century, and it is likely that the resultant general lowering of the water table has caused the dessication and disintegration of any other boat remains in the ground, while the extensive agricultural practices characteristic of the lower-lying parts of the region are not conducive to the discovery and recognition of archaeological timberwork. There are probably few further discoveries to be made.

*IV.2.5. TAY ESTUARY*

(8 discoveries)

| | | |
|---|---|---|
| 7 | Barry Links | c. 1820 |
| 37 | Errol 1 | c. 1869 |
| 38 | Errol 2 | 1895 |
| 50 | Friarton | 1878-9 |
| 87-8 | Lindores 1-2 | c. 1816 |
| 151 | 'River Tay' | undated |
| 152 | Sleepless Inch | 1848 |

The distributions of logboat discoveries around the estuaries of the Tay and the Forth are essentially similar, and the two are considered together in the next section (IV.2.6).

*IV.2.6. FORTH VALLEY AND ESTUARY*

(7 discoveries)

| | | |
|---|---|---|
| 14 | Cambuskenneth | 1874 |
| 45 | Falkirk | pre-1805 |
| 142-3 | Port Laing 1-2 | c. 1857 |
| 148 | River Carron | 1726 |
| 150 | River Forth | c. 1836-9 |
| 153 | Stirling, King Street | 1865 |

The patterns of discovery in the Tay and Forth estuaries are the same as that noted for the Clyde (section IV.2.2, above) except that no discoveries have been made in either of the East coast rivers during the present century, and that more discoveries have been made in river-dredging operations than in port construction works. It is highly probable that numerous boats, some of them possibly of very early date, remain to be discovered, both in estuarine sandbanks and in the river terraces. The commercial use of Grangemouth and Perth will necessitate continuing dredging, and possibly also deepening, as far as these ports, while continuing urban development in the surrounding areas may reveal further examples.

*IV.2.7. MISCELLANEOUS*

(11 discoveries)

| | | |
|---|---|---|
| 19 | Castlemilk | c. 1831 |
| 22 | Clune Hill, Lochore | 1926 |
| 47 | Flanders Moss | late 18th century |
| 52 | Gartcosh House | c. 1892 |
| 81 | Kinross | c. 1862 |
| 85 | Lea Shun | 1887 |
| 86 | Lendrick Muir | probably 1881 |

| | | |
|---|---|---|
| 89 | Linlithgow, Sheriff Court-house | c. 1860 |
| 90 | Littlehill | 1870 |
| 116 | Loch Leven | probably 1830-2 |
| 140 | 'Orkney' | undated |

The heterogenous discoveries noted in this section serve to emphasise that logboats may be discovered anywhere in Scotland, and have probably been one of the types of artefact used most widely across the country. Until demonstrated otherwise, it must be assumed that the apparent gaps in the recorded distribution that have been noted in the eastern part of Lothian, in Berwickshire and the eastern Borders, and in the far north are the result of incomplete recording and that examples await discovery in these areas.

## IV.3 Surviving Examples and Detailed Records

Twenty-eight Scottish examples survive (at least in part) out of 150 recorded discoveries, yielding a survival percentage of 19% which is noticeably worse than the comparable figures of 179 discoveries, 75 surviving examples and a surviving percentage of 42% for England and Wales. This discrepancy is probably to be attributed as much to the relative scarcity of Scottish local museums and storage facilities as to the relatively early date of recorded Scottish discoveries, a high proportion of which date from the eighteenth and early nineteenth centuries, and comparatively few from the inter- and post-war periods. The notably low percentage of survivals in the Forth valley area, where archaeological research has traditionally been most intense and where many early discoveries were made, suggests that the second factor was the more important.

The surviving examples may usefully be divided into the following categories according to their states of preservation:

1. Essentially complete; both ends survive, the full width of the beam survives amidships, and both sides survive, at least in part, to their full height,

2. The complete length of the bottom of the boat survives, but only the lower parts of the sides,

3. One or both ends of the boat are missing, and the sides are only partly preserved,

4. The boat has been reduced to a flat plank, and

5. Only fragments survive, some of which may not be identifiable as specific portions of the boat.

On this basis, the summary figures for Scotland as a whole are cited, with comparative figures for England and Wales, in table 2.

The Scottish survivals may be similarly analysed along the geographical lines that were defined in section IV.2 to yield a reasonably even pattern across the country as a whole, about 30% of the examples discovered having been recovered and still extant, while a few more survive *in situ*. Possibly surprisingly, the percentage survivals are nearly consistent across the country, with little reflection being evident of the existence of major museum collections in Edinburgh, Glasgow and Dumfries.

Detailed figures and percentages for specific areas are cited in table 3. If subdivided on the basis of the preservation categories defined above, the surviving examples may be divided as cited in table 4. Again, no significant variation appears between regional groups, apart from the significant number of examples remaining *in situ* in highland lochs and bogs.

## IV.4 Structural Features

The methodology laid down by McGrail (1978, *passim*.) for logboat analysis was devised on the basis of the study of the specimens preserved in English and Welsh museums; these early discoveries were nearly-invariably inadequately recorded and had in many cases suffered from incorrect conservation and storage. Subsequent practical recovery operations have allowed the methodology to be refined, particularly in respect of stability calculation formulae (McGrail 1987a, 57-87 and *pers. comm.*).

Further refinement has been made possible by the detailed study and extensive publication (Millett and McGrail 1987) of the logboat that was discovered in 1984 at Hasholme, Holme-on-Spalding Moor in the lower Vale of York. Although this Iron Age vessel is of a size and complexity which are apparently rarely paralleled, a detailed study of her construction serves to indicate the range of variation that are possible in this class of artefact, while vessels of her type can be envisaged as contributing substantially to the viability of a developed rural and industrial society in an area which is topographically not dissimilar from parts of lowland Scotland.

This boat was remarkable for her size; at about 12.6m (including two detached bow-timbers) she is probably the second longest recorded in England, but some 1.1m shorter than the longest Scottish example (Loch Arthur 1), the beam measurements being roughly comparable. The estimated light displacement of 4.4 tonnes (McGrail 1988, 44) is a guide to the impressive size of the larger logboats, an aspect of their design which may have had a considerable prestige value.

The Hasholme boat was also remarkable for her state of preservation, even though the sides had been distorted by post-depositional compression and the bow was seriously damaged before recovery; the upper parts of the sides and most of the fittings were recovered for exami-

| State | Scotland | England and Wales |
|-------|----------|-------------------|
| 1 | 4 (14%) | 21 (28%) |
| 2 | 8 (29%) | 3 (4%) |
| 3 | 7 (24%) | 31 (41%) |
| 4 | 7 (24%) | Nil |
| 5 | 3 (10%) | 20 (27%) |
| Total | 29 | 75 |

*Table 2. Comparative figures for states of logboat preservation in Scotland as against England and Wales, in accordance with the criteria defined in section IV.3.*

nation. The hull had apparently been formed, presumably in the normal sequence, with the use of metal tools but without that of fire; fortuitous assistance was provided by the extensive heart-rot in the timber and the builders were guided by the provision of seven or eight thickness-gauge holes. The workmanship was of a commendably high quality, with typical thicknesses of 0.25m and 0.09m being attained for the bottom and sides respectively. Beam-ties had been fitted high up amidships and aft of the transom, and were retained by interference fits, trenails and cottered joints.

Although the bow was seriously damaged during discovery, enough remains to demonstrate that its form does not fall into the normal logboat pattern. It can be shown to have comprised a two-piece trenailed structure incorporating a substantial bow-platform; this demonstrates the need for a distinction between the concepts of 'side-extension' and 'end-extension' (Millett and McGrail 1987, 115); the latter feature is previously-unrecognised and has no parallel in Scotland. Nor is there any evidence on any boat from Scottish antiquity for a cleat (worked from the solid) of the type that forms part of the lower bow of the Hasholme boat and is fundamental to the construction of such boats as the Bronze Age vessels from Ferriby and Brigg in and around the Humber estuary. Side-extension, which is evidenced in Scotland by the examples from Buston (no. 11), Dowalton Loch 1 (no. 29), the 'River Clyde' (no. 149) and, also, the possible and partial examples from Erskine 1 (no. 39) and Lochmaben, Castle Loch 2 (no. 131), is indicated by the presence of a trenailed-retained washstrake, which measures between 3.7m and 4m in length, for the starboard side, the former presence of an equivalent feature on the port side being inferred.

Damage and repairs are an inevitable consequence of small boat operations, and the Hasholme vessel bears evidence of repair at two points. The repair-patch (measuring some 1.25m in length) that has been rabbeted onto the upper part of the port side and was probably retained by trenails, and the trapezoidal block (measuring some 0.72m in length) that was inserted into the inner face of the starboard side across the transom-groove and there retained by a horizontal trenail, represent two approaches to the problem that are paralleled in Scotland, the first most probably at Buston 1 (no. 11) and the second, possibly, at Lea Shun (no. 85). The repair to the stern of the Hasholme boat was apparently carried out to repair, or at least contain, a split which was caulked with moss, a practice which may be evidenced in Scotland if caulking was inserted beneath the over-sewn batten used to repair the Loch Laggan 2 boat (no. 110). There is no evident Scottish parallel for the placing, as at Hasholme, of birch or softwood inserts or graving-pieces within the gaps in one of the larger knots.

The features noted at Hasholme provide a possible type-specimen of a developed eastern English logboat, which owes much to the complex cleat-built construction of the Ferriby boats and the Brigg 'raft'. The lack of Scottish parallels for many of the features noted may be the result of a poor survival rate, or may indicate that many of the Scottish boats are of a later date.

Those 'modern' morphological features that may indicate the derivation or skewomorphic copying of specific examples from plank-built boats are discussed below (section IV.7.5).

## IV.5 Size, Form and Morphology

The morphological analysis of Scottish logboats may be placed in perspective by a consideration of the possibility of recognising dated local groups in certain areas. Such a form of analysis requires a considerable body of evidence and is constrained (particularly in its statistical aspects) by the inflexibility of logboat construction within the constraints of the available timber size and type; the size-grouping that is defined may be effectively that of the available trunks. Also, classification may not, in practice, depend on the recognition of minor features (particularly those found high in the sides) as these will survive only infrequently. In practice, the recognition of regional groupings is likely to prove an unsophisticated analytical tool.

Nevertheless, such an exercise has been attempted with some success. Notably, Joncheray (1986) has considered the 25 examples from the Loire basin in western France. These apparently range in date from the fourth century bc to the eleventh century ad, and exhibit some distinct regional differences. So far as can be determined, the Loire examples are all relatively small, have both ends in the form of rounded points, and have been worked with few (if any) thickness-gauge holes. None of them have transoms, keels, or other possibly skewomorphic features, although several have transverse ridges left in the solid, one has a marked longitudinal projection along the centreline and another had a seat left in the solid at each end.

| Area | Discoveries | Survivals | Percentage of regional discoveries |
|---|---|---|---|
| Dumfries, Galloway and Ayrshire | 52 | 13 | 25% |
| Clyde estuary and river-terraces | 34 | 8 | 24% |
| Highland lochs and bogs | 27 | 10 | 37% |
| Strathmore, Angus, the Mearns and the North-east | 14 | 4 | 29% |
| Tay estuary | 8 | 2 | 25% |
| Forth valley and estuary | 7 | 1 | 14% |
| Miscellaneous | 10 | 3 | 30% |
| **Total** | **154** | **41** | **27%** |

*Table 3. Comparative figures for logboat preservation in specific areas of Scotland.*

Joncheray further attempts to sub-group the Loire examples according to their size and form, isolating one group of seven examples found in a limited geographical distribution along a 45-km length of the river around Nantes, including the example from Ancenis, le Pont which is radiocarbon-dated to 940 ± 60 ad. This group is characterised by straight, parallel and vertical sides, flat bottoms and ogival ends as well as a rough similarity of dimensions at about 5m in length and between 0.5m and 0.9m in beam. In only one case can the elevation be established, and in general there are variations in the form of the ends which can be best attributed to the constraints of the available timber.

In a significantly different geographical context, the evidence for a distinct group of logboats in Denmark has been considered and summarised by Christensen (1990) who notes the discovery of about 250 logboats and over thirty paddles in the country. Around 50 of these boats can be attributed to the Stone Age, these being typically of softwood construction, longer than later vessels and of rounded cross-section. Mesolithic boats being all of lime and found in coastal contexts, while the Neolithic vessels are more typically found inland and of alder. In neither period was fire used in their construction, but hearths laid on beds of clay have often been noted, and both ballast stones and evidence for repair were noted. Both solid and 'open' sterns were noted, the latter being closed by transoms or clay plates.

A chronologically-based synthesis of the logboats of Northern France and the Ile de France, where both hardwood and softwood traditions appear to have existed would be a most useful exercise.

An interesting and relevant study from another continent is that by Plane (1991) who identifies two loosely-defined types (for coastal marine and inland use, respectively) in northern New England, New Brunswick and Quebec. Of thirty-three recorded vessels, only four can be shown (by radiocarbon assay) to pre-date the period of European colonisation, and these may be typologically distinguished (by their rounded form) from the 'Euro-American' type which displays a wide range of forms but typically has a pointed bow and a square or truncated stern. Ancillary features commonly found include 'pegs' (possibly thickness-gauges), nails, 'seats', 'mast braces' and evidence for repairs. These discoveries are of particular interest for being found in an area where birch bark vessels of various types have generally been considered prevalent.

Early European travellers in the region described the manufacture of logboats from (usually) pine or chestnut with the use of fire, stone axes, heated pebbles, stone scrapers and clam shells. One possible logboat-manufacturing site has been recognised in excavation and tentatively dated to the middle or late Archaic Period (c. 5000-1000 BC).

During the early years of colonisation, Europeans also built their own 'Cannows', possibly under the influence of both native and European traditions. Trees intended for logboat manufacture were, in some areas, specifically protected in law. Such vessels remained in general use into the eighteenth century, their manufacture using thickness-gauges of red cedar is recorded in the late nineteenth century, and they remained in occasional use in places into recent times.

More locally, the characteristics of those English and Welsh logboats known to 1978 have been drawn together by McGrail (1978, i, 309-27) and the Scottish examples may usefully be compared with them in terms of type of timber used, dimensions and form. On the basis of accepting the traditional, and now often unverifiable, identification of timber types, the comparable figures are cited in table 5. Given the relatively small sample involved and the probable unreliability of many of the identifications, no significant pattern can be inferred from these figures.

| Regional Group | 1 | 2 | 3 | 4 | 5 | In situ |
|---|---|---|---|---|---|---|
| Dumfries, Galloway and Ayrshire | 2 | 3 | 3 | 4 | - | 1 |
| Clyde estuary and river-terraces | - | 3 | 1 | 3 | - | 1 |
| Highland lochs and bogs | 1 | - | 2 | 2 | 1 | 6 |
| Strathmore, Angus, the Mearns and the North-east | 1 | 1 | - | - | 2 | - |
| Tay estuary | 1 | - | - | - | 1 | - |
| Forth valley and estuary | - | 1 | - | - | - | - |
| Miscellaneous | 1 | 1 | - | - | 1 | - |
| **Total** | **6** | **9** | **6** | **9** | **5** | **8** |
| **Percentage of total sample (43)** | **14** | **21** | **14** | **21** | **12** | **19** |

*Table 4. comparative figures for state of surviving logboats in specific areas of Scotland.*

However, the dimensions of the recorded Scottish logboats are notably smaller than those found in England and Wales, the mean lengths of recorded examples measuring 5.16 ± 2.35m in England and Wales as against 4.71 ± 3.0m in Scotland. The difference in beam is predictably less significant, the Scottish examples measuring 0.9 ± 0.31m as against 0.87 ± 0.29m for those from England and Wales. The percentages of recorded boats of different (minimum) lengths are set out as table 6. This indicates the slightly shorter average length found in Scotland, as a result of the recorded discovery in England and Wales of only a few small fragments and the existence of such very long boats as those from Brigg, Lincolnshire (14.78m), Hasholme, Yorkshire (12.78m) and Poole, Dorset (10.01m). The presumed smaller size and stunted growth of Scottish oaks (near the margin of the distribution of the species) cannot be unequivocally shown to have had an effect but must also be considered to have been a factor.

McGrail divides the recorded English and Welsh logboats into morphological categories along the lines noted above and these are compared quantitatively with the Scottish examples in table 7. The greater prevalence of dissimilar-ended forms in Scotland is at once apparent. Of the 36 examples that have been noted, 12 (26%) have solid sterns while 24 (51%) have transoms; comparable figures for the fifteen recorded examples from England and Wales are 13 boats and 87% respectively. However, these figures may be inherently distorted to exaggerate the number of dissimilar-ended examples; to demonstrate the presence of a transom it is only necessary to recover the stern, or sufficient of it to reveal a transom-groove, whereas the existence of two similar (solid) ends can only be demonstrated by the recovery of practically the entire boat. In practice, the specific morphology codes that McGrail proposes for each form are not well-matched in Scotland, and most of the Scottish boats are of variant form.

These observations may possibly be related to the higher percentage of side-extended logboats that has been noted in Scotland. Only one undoubted example (Kentmere 1) is recorded in England or Wales but the following Scottish examples of side-extension have been noted:

11   Buston 1
29   Dowalton Loch 1
39   Erskine 1 (partial example)
131  Lochmaben, Castle Loch 2 (partial example)
149  'River Clyde'

There are neither known examples of end-extension nor recorded examples of expanded logboats in Scotland, although the latter mode of construction has been implausibly claimed for Errol 2 (no. 38). The apparent absence of large aspen trunks in sufficient quantities makes it unlikely that such construction practices were as prevalent as has been noted in Finland (NMM 1985, plates 15-17).

Taken together, these observations may possibly be related to the presumed reduced availability of large oak trunks in Scotland and the consequent need to adopt methods of construction which utilise timber of smaller dimensions.

## IV.6 Repairs

As the evidence of the Hasholme logboat demonstrates, minor accidents and the subsequent necessary repairs are an inevitable result of logboat use, while storage out of water for any length of time usually results in the timber drying with consequent longitudinal and radial splitting, and possibly splitting in the more extreme cases. Indeed, Earwood (*pers. comm.*) has pointed out that many 're-

|  | England and Wales | | Scotland | |
| --- | --- | --- | --- | --- |
|  | Number | % | Number | % |
| Oak | 76 | 96 | 11 | 87 |
| Pine | 1 | 1.3 | Nil |  |
| Elm | 1 | 1.3 | Nil |  |
| Ash | 1 | 1.3 | Nil |  |
| Fir | Nil |  | 5 | 13 |

*Table 5. Figures for types of timber identified in Scottish logboats with comparable figures for England and Wales.*

pairs' may have been carried out during construction to remedy defects in the available timber.

It is inherent in the manufacture of logboats by a reduction technique that new components cannot be fabricated to directly replace those rendered defective, but, rather, that repair-patches or lashings must be fabricated either from a similar material to the rest of the boat or from such dissimilar materials as iron or lead, which may be caulked with moss and retained in place br wooden trenails, oversewn fastenings of natural vegetation, or iron pegs and nails. Ethnographically-recorded modern practices include the use of nail-retained sections of truck tyre in Cameroon (Dodwell 1989, 18).

Ten of the recorded Scottish logboats display evidence for repairs in a variety of forms:

### IV.6.1. REPAIR BY PATCH NAILED OR PEGGED OVER HOLE OR KNOT

11  Buston 1
Two repair-patches, presumably of wood, retained by 'ribs'.

54  Glasgow, Clydehaugh 2
Possible example: lead patch found beneath stem, but no evidence on boat.

57  Glasgow, Clydehaugh 5
Wooden patch fastened externally.

75  Kilbirnie Loch 1
Possible example: 'thin plate of metal' found in bottom of boat.

### IV.6.2. BLOCK OR TIMBER INSERTED INTO HOLE OR SPLIT

35  Dumbuck
Wooden 'clamps' along split.

39  Erskine 1
Double-flanged timber insert measuring 230mm long.

85  Lea Shun
Possible example: two 'oak' patches inserted and fastened into place in knots.

105  Loch Kinord 1
Two 'bars of oak' retained by five dovetail-jointed transverse bars.

### IV.6.3. OVERSEWN BATTEN ALONG SPLIT

110  Loch Laggan 2
Two battens along longitudinal split.

Repairs made using any of these methods could be supplemented by additional moss-caulking, although this is only directly evidenced in the case of Buston 1.

## IV.7 Dating and Chronology

The conventional view on the dating of Scottish logboats is best summarised in the guide-catalogue of the Hunterian Museum, University of Glasgow (Robertson 1954, 7). They are noted in the Neolithic section but the remark is made that 'Although tree-trunk canoes had come into use at least as early as the Late Stone Age, their manufacture continued into very much later times'.

This view is derived from the work of Buchanan (1848) and Wilson (1863, i, 50-8) who recorded discoveries in carseland, river-terrace and lake-bed deposits and stressed their demonstrably high antiquity in a period when the French Palaeolithic discoveries of Boucher de Perthes (in comparable geologically-stratified contexts in the Somme valley gravels) were attracting great interest. At least one attempt was made (Lyell 1829) to reverse the argument and support the early dating of a lake-bed deposit by the discovery of a logboat in it. Although some modification of this view was required to extend the lifespan of this class of artefact into later prehistory on the evidence of discoveries in and around crannogs, this is apparently the view generally held in the earlier part of this century. The

## IV Synthesis and Analysis

|  | England and Wales | | Scotland | |
|---|---|---|---|---|
|  | Number | % | Number | % |
| 1-1.99m | Nil |  | 3 | 3 |
| 2-2.99m | 8 | 11 | 8 | 9 |
| 3-3.99m | 19 | 25 | 22 | 25 |
| 4-4.99m | 15 | 20 | 18 | 21 |
| 5-5.99m | 9 | 12 | 7 | 8 |
| 6-6.99m | 5 | 7 | 11 | 13 |
| 7-7.99m | 4 | 5 | 9 | 10 |
| 8-8.99m | 3 | 4 | 3 | 3 |
| 9-9.99m | 4 | 5 | 2 | 2 |
| 10-10.99m | 4 | 5 | 1 | 1 |
| 11-11.99m | 1 | 1 | 2 | 2 |
| 12-12.99m | 1 | 1 | Nil |  |
| 13-13.99m | Nil |  | 1 | 1 |
| 14-14.99m | 2 | 3 | Nil |  |
| 15-15.99m | Nil |  | Nil |  |
| 16-16.99m | 1 | 1 | Nil |  |
| **Total** | **76** |  | **87** |  |

*Table 6. Figures for recorded lengths of Scottish logboats with comparable figures for England and Wales.*

|  | England and Wales | | Scotland | |
|---|---|---|---|---|
|  | Number | % | Number | % |
| Canoe | 19 | 36 | 8 | 17 |
| Punt/barge | 15 | 29 | 3 | 6 |
| Dissimilar ends | 15 | 29 | 36 | 77 |
| Box | 3 | 6 | Nil |  |
| **Total** | **52** |  | **47** |  |

*Table 7. Figures for identifiable forms of Scottish logboats with comparable figures for England and Wales.*

parallel of some aspects of construction of the Clyde logboats in more modern ships was noted but its chronological significance was not appreciated.

The best available parallel for the use of logboats in Mesolithic Scotland is provided by the underwater excavation of part of a settlement of the Ertebolle culture together with its cemetery, refuse area and littoral fishing ground at Tybrind Vig, Fyn, Denmark (Andersen 1985; 1986). The settlement was occupied throughout greater part of both the aceramic (pre-3700 bc) and pottery-using phases of that late Mesolithic culture, between roughly 4600 and 3200 bc. Among the structures there was recognised a probable dock comprising 'thick posts and a cobbled area' while discoveries of leister prongs, bone fish-hooks and the remains of probable fish-traps confirm the bone and carbon-13 evidence for a local economy which extensively exploited cod, spurdog and eel in the latter phases of occupation. Some of the numerous textile and rope objects discovered presumably had maritime uses.

The two logboats that were found in the upper levels of the site were of trough form and worked from straight trunks of lime with well-shaped and smoothed sides at least 300mm high. Both bore extensive toolmarks which indicated manufacture by adze-chopping alone, apparently without longitudinal cleavage or the use of fire. Indentations between 20 and 400mm wide, and consistent with the use of an Ertebolle-type flint axe or adze, were found extending diagonally from the upper sides to the bottoms of the boats, where there were prominent longitudinal marks of apparently-similar origin. Fragments of further, possibly uncompleted, boats were also found, and seen as evidence of the known tendency of lime to shatter if not carefully worked. It is estimated that two people might make such a logboat by adze-work within about a week, leaving only a concentration of waste flakes and flint adzes for the archaeological record. Boat 1, which was dated by radiocarbon to 3310 ± 95 bc (K-3557), was found complete within the 'waste layer' in the reed swamp immediately outside the settlement. It had apparently been U-sectioned and measured 9.5m in length by about 0.65m in beam; the sides were between 200 and 300mm thick and the square-cut stern was pierced by seven holes which has served to retain a (now-lost) transom. Andersen (1986, 90) calculates that the boat was worked from the trunk of a tree between 100 and 150 years old, which measured between 12 and 14m in length by between 0.8 and 1m in diameter, and had been worked (to avoid splitting) by forming the bow from the root and the stern from the crown. Within the boat which had an estimated carrying capacity of between six and eight persons, there were found a probable oval fireplace of sand and small stones, and a possible ballast-stone.

Only the transom-fitted stern portion of the second boat was found in similar stratigraphic conditions. This boat, which had apparently been larger, was dated to 3420 ± 95 bc (K-4149). Traces of a possible fireplace were similarly found in this boat and, like that in boat 1, were interpreted as an 'eel flare' for use in fishing. Evidence for two repair-patches was found, one of them across a mid-axis split and the other across a knot.

In England, the realisation that many examples might be of medieval date was brought about by the publication of the extended logboat dated by radiocarbon to 1300 ± 135 ad from Kentmere (Wilson 1966; McGrail 1978, i, 223-5, no. 70; Booth 1984, 193) in the English Lake District, where Fox (1926, 128; McGrail 1978, i, 300) had previously reported a (possibly spurious) local tradition of logboat usage into the earlier part of the 19th century. This dating has since been confirmed by the attribution of a radiocarbon date of 1335 ± 40 ad (Q-1245) to the boat of similar construction form Giggleswick Tarn, Yorkshire (McGrail 1978, i, 190-5, no. 49; Booth 1984, 192) and the closely-spaced dating in the 9th to 12th centuries ad of those from the River Mersey at Warrington (McGrail and Switsur 1979a). Current orthodoxy in England is best expressed by McGrail and Switsur (1975; 1979) who stress the frequency of medieval examples, but cite none later than the 14th century.

There is no evident reason why the use of logboats should not have continued in Scotland to a later date than this, as it did in Ireland (Lucas 1963) where such boats were evidently used in inland waters rather than the better-known skin-built currach which was preferred for seagoing and coastal use. The Old Irish words 'coite', 'cott' and 'cot' are distinguished from other types of boat and identified with the logboat while a 'crand lestra' (literally 'tree vessel') is also mentioned in the Irish Laws. The term 'trough' is also used on occasion in a logical but confusing manner.

Numerous Irish accounts written in the 14th to 17th centuries mention the use of logboats by raiding parties (particularly against crannogs) as well as in more orthodox military operations. Some information is given regarding the size of the vessels; an entry in the Annals for 1505 relates the loss of eighteen persons in an accident involving one cot on Lower Lough Erne, while in 1593 the 'drowning' (which is presumably different from sinking) is recorded of several cots, each of them capable of carrying ten men. A further account, which is not noted by Lucas, describes (O'Morain 1957, 51) the use of a seagoing logboat to aid the escape of the former Prior of Kilkenny from the Cromwellian attack on Burrishoole Convent, Co. Connaught. He is recorded as writing 'I myself, with one boy, managed to get a dugout canoe and in that tiny boat I launched into the deep...I made Clare Island in safety. It was a journey of six leagues on the high seas'. By way of illustration of other Irish uses for the same woodworking techniques, in the late 17th century John Dunton (cited by Earwood 1993, 237) mentions the working of 'a square wooden vessel, called a meddar all of one piece cutt out of a tree'.

Although there is no unambiguous reference to logboat use in Ireland after 1698, it is extremely possible that such boats remained in use until the late 18th century on the extensive inland loughs. Lucas (1963, 66) suggests that their construction declined for want of suitable trees after the extensive felling of the Irish oak forests to obtain raw material for the export trade in barrel staves.

Similar, but not identical, factors were probably important in Scotland where logboat construction was probably already constrained by the absence of extensive tracts of deciduous, and specifically oak, woodland across most of the North and West of the country. In Argyll, which might be expected to have been a major area of logboat use on account of its climatically-determined woodland, the rapid destruction of this resource to produce iron-working charcoal in the late 18th century parallels the Irish deforestation.

There are few references to boats of any description in the various accounts of the Jacobite risings and in the travellers' tales that were written after 1746. Burt mentions (Jamieson 1876, ii, 42-3) only a boat of unspecified type and location which was 'above sixty years old... patched almost everywhere with rough Pieces of Boards, and the Oars were kept in their Places by small Bands of twisted sticks' while Johnson notes that the boats of the Western Isles and Lochaber were not 'furnished with benches, or made commodious by any addition to the first fabric' (Chapman 1924, 93), but in neither case is there any unequivocal statement that the account is of anything other than a wooden boat of conventional type. What is probably an account of a pair of conjoined logboats dates from 1760 when Pococke (1887, i, 37) passed through Annandale and saw 'a double kind of boat, like two troughs joined...each of which would hold any beast to be carried over'.

This account appears to corroborate the suggestion by Osler (1985) that the Northumbrian trow was possibly derived from such a linked pair of logboats. This type of double-hulled boat was traditional to the salmon-fisheries of the Northumbrian rivers and measured about 10' (3.05m) in length. The single occupant (if that is the correct term) propelled the boat by punting while standing with one foot in each of the two hulls; these were slightly-built of deal with numerous transverse strengthening-pieces and each measured about 1'2' (0.36m) in beam by 1' (0.3m) 'deep'. The hulls were joined directly at the bow and by a flat board at the stern to give the boat a triangular form in plan view. Although no evident examples of paired logboats have been identified in Scotland, the use of such a type would make good use of the small-dimensioned timber that was probably most widely available in Scotland, and may also explain the diminutive size of some Scottish examples.

Joass (1881) describes two types of boats formerly used in the Highlands on the basis of a letter written in 1798 by the then Minister of Dornoch, Sutherland. The 'courich' was a small round wicker-framed boat covered with 'green hides' built to carry two or three people and formerly used as a ferry on rivers and small creeks. In Wester Ross, this had been superseded by the 'ammir' or 'trough', which was 'a sort of Canoe...nothing more than the hollowed Trunk of a great Tree'. The ammir was said to have been propelled with surprising speed by a standing man holding an 'oar' (probably a double-bladed paddle) in the middle so as to stroke on each side alternately.

There seems no reason to doubt that this pattern is common across Europe, and that logboats are by no means universally of Mesolithic date. The discovery of a model logboat in an unstratified context at the Linear Pottery Culture settlement site of Wiesbaden-Erbenheim in the Rhine-Main area of West Germany (Peschel 1985) may be noted.

For comparison, the available radiocarbon dates for non-Scottish European logboats are tabulated (in order of increasing antiquity) as table 8. Against this background, the evidence for the various dated logboats of Scotland is here tabulated according to the manner of their dating, with brief notes on each category.

*IV.7.1 RADIOCARBON DATES*

(tabulated in reverse chronological order)

49   Forfar 2
$1090 \pm 50$ ad   Cambridge

118   Loch of Kinnordy
$735 \pm 40$ ad   Cambridge

96   Loch Doon 1
$509 \pm 110$ ad   SRR-501

38   Errol 2
$485 \pm 40$ ad   Cambridge

44   Erskine 6
$45 \pm 50$ bc   GU-1016

92   Loch Arthur 1
$101 \pm 80$ bc   SRR-403

20   Catherinefield
$1804 \pm 125$ bc   SRR-326

The number of logboats, and of portable wooden antiquities in general, that have been subjected to radiocarbon assay is not high, but it serves to emphasise the wide range of dates involved. It appears reasonable to hope that the decreasing cost of laboratory determinations will soon allow the dating of all the surviving boats and any further discoveries.

*IV.7.2. ARCHAEOLOGICAL ASSOCIATION, INCLUDING INCORPORATION INTO CRANNOG STRUCTURES*

29-30         Dowalton Loch 1-2

| Name and location | Radiocarbon determination | Laboratory number |
|---|---|---|
| Slavonski Brod, River Sava, Yugoslavia | 1710 ± 80 ad | Z-553 |
| River Sava, Yugoslavia | 1700 ± 100 ad | BC-42 |
| Arslev Edge, Moesgard, Denmark | 1466 ± ?? ad | K-1213 |
| Walthamstow, London, England | 1604 ± 54 ad | BM-961 |
| Weybridge, Surrey, England | 1540 ± 60 ad | Har-4996 |
| Lac de Paladru, Lepin, Isère, France | 1370 ± 270 ad | Ly-2274 |
| Giggleswick Tarn, Yorkshire, England | 1335 ± 40 ad | Q-1245 |
| Kentmere 1, Westmorland, England | 1300 ± 120 ad | D-71 |
| Moen, Solum, Telemark, Norway | 1210 ± 110 ad | T-1429 |
| Trasimeno Lake, Italy | 1206 ± 100 ad | Pi-84 |
| Warrington 1, Lancashire, England | 1190 ± 60 ad | Q-1390 |
| Warrington 7, Lancashire, England | 1090 ± 60 ad | Q-1395 |
| Irlam, Lancashire, England | 1085 ± 40 ad | Q-1456 |
| Warrington 4, Lancashire, England | 1072 ± 60 ad | Q-1393 |
| Le Cellier, La Saulzaie, Loire-Atlantique, France | 1070 ± 60 ad | Gif-7040 |
| Garnat sur Engièvre, Allier, France | 1050 ± 110 ad | Ly-2252 |
| Barton, Lancashire, England | 1030 ± 65 ad | Q-1396 |
| Warrington 2, Lancashire, England | 1020 ± 50 ad | Q-1391 |
| Monate 1, Varese, Lombardy, Italy | 1010 ± 75 ad | F-62 |
| Warrington 11, Lancashire, England | 1000 ± 90 ad | Birm-269 |
| Stanley Ferry, Yorkshire, England | 990 ± 70 ad | Har-2835 |
| Warrington 5, Lancashire, England | 958 ± 65 ad | Q-1394 |
| Ancenis, La Davrays, Loire-Atlantique, France | 940 ± 60 ad | Gif-7041 |
| Warrington 3, Lancashire, England | 875 ± 60 ad | Q-1392 |
| Peschanoye, Gel'myazev Raion, Cherkassy, Ukraine | 830 ± 100 ad | LE-654 |
| Llyn Llangorse, Brecon, Wales | 814 ± 60 ad | Q-857 |
| Haukasmyra, Asvang, Stange, Hedmark, Norway | 810 ± 80 ad | T-2052 |
| Sainte-Anne, Le My, Loire-Atlantique, France | 760 ± 60 ad | Gif-5430 |
| Epervans, Soane 1, Seine et Loire, France | 690 ± 140 ad | Ly-2199 |
| Amberley 3, Sussex, England | 640 ± 70 ad | Q-828 |
| Flavigny-sur-Meuse, Meurthe-et-Moselle, France | 540 ± 80 ad | (unstated) |
| Mattersea Thorpe, Lincolnshire, England | 460 ± 80 ad | Har-4997 |
| Parc de la Tête d'Or, Lyon, Rhône, France | 450 ± 110 ad | Ly-68 |
| Monate 2, Varese, Lombardy, Italy | 370 ± 105 ad | F-63 |
| Hardham 2, Sussex, England | 295 ± 50 ad | Q-827 |
| Chaudeney sur Moselle, Toulouse, France | 200 ± 70 ad | Ny-314 |
| Chaudeney sur Moselle, Toulouse, France | 100 ± 60 ad | (unstated) |
| Egernsund, Denmark | 10 ± 75 bc | K-2513 |
| Egernsund, Denmark | 50 ± 75 bc | K-2514 |
| Jyllinge, Denmark | 90 ± 75 bc | K-2898 |
| Ancenis, le pont, Loire-Atlantique, France | 130 ± 150 bc | Gsy-236 |

*Table 8. Summary of recorded radiocarbon dates for European logboats in reverse chronological order.*

| Name and location | Radiocarbon determination | Laboratory number |
|---|---|---|
| Holme Pierrepont 1, Nottinghamshire, England | 230 ± 110 bc | Birm-132 |
| Lassby, Sweden | 265 ± 110 bc | St-3550 |
| Poole, Dorset, England | 295 ± 50 bc | Q-821 |
| Shapwick, Somerset, England | 355 ± 120 bc | Q-357 |
| Oudon, L'ile Neuve, Loire-Atlantique, France | 370 ± 60 bc | Gif-5431 |
| Hasholme, Holme-on-Spalding Moor, Humberside, England | 330 ± 80 bc<br>400 ± 90 bc<br>600 ± 100 bc | Har-6441<br>Har-6394<br>Har-6395 |
| Cret de Châtillon, Sevrier, Haute-Savoie, France | 750 ± 140 bc | Ly-1951 |
| Brigg, Lincolnshire, England | 814 ± 100 bc | Q-78 |
| Skaggered, Sweden | 835 ± 100 bc | St-3551 |
| Appleby, Lincolnshire, England | 1100 ± 80 bc | Q-80 |
| Berliner See, Berlin, Germany | 1160 ± 60 bc | (unstated) |
| Lago Lucone, Brescia, Lombardy, Italy | 1210 ± ?? bc<br>1410 ± ?? bc | R-375a<br>R-375 |
| Chapel Flat Dyke, Yorkshire, England | 1500 ± 150 bc | BM-213 |
| Bande de Cavriana, Mantua, Lombardy, Italy | 1570 ± 50 bc | R-786a |
| Branthwaite, Cumberland, England | 1570 ± 10 bc | Q-288 |
| Meimart, Le Bourget Lake, Brisson-Saint-Innocent, Savoie, France | 1790 ± 130 bc | Ly-2305 |
| Les Baigneurs, (Cha 4), Pirogue A, Lac Paladru, Isère, France | 2240 ± 150 bc | Ly-792 |
| Verup-komplekset boat 1, Denmark | 2270 ± 85 bc | K-4098B |
| Ogarde-komplekset boat V, Denmark | 2330 ± 85 bc | K-3637 |
| Praestelyngen BII, 378, W Zealand, Denmark | 2470 ± 110 bc | K-1649 |
| Kildegard-komplekset boat II, Denmark | 2550 ± 85 bc | K-4338 |
| Bolling so III, Denmark | 2560 ± 120 bc | K-1214 |
| Sondersted boat I, Denmark | 2590 ± 90 bc | K-3638 |
| Lago di Fimon, Val di Marcia, Vicenza, Veneto, Italy | 2630 ± 50 bc | R-359a |
| Ogarde-komplekset boat III, Denmark | 2640 ± 120 bc | K-1165 |
| Bodal-komplekset boat II, Denmark | 2740 ± 110 bc | K-2177 |
| Brokso, Holmegards Mose, Denmark | 2840 ± 90 bc | K-4099 |
| Praestelyngen BII, 421, 428, 447, W Zealand, Denmark | 2940 ± 110 bc | K-1651 |
| Praestelyngen boat II, W Zealand, Denmark | 2980 ± 75 bc | K-2009 |
| Praestelyngen BII, 128, W Zealand, Denmark | 3010 ± 110 bc | K-1650 |
| Praestelyngen BII, 1, W Zealand, Denmark | 3060 ± 100 bc | K-1473 |
| Flynderhage, Denmark | 3280 ± 100 bc | K-1450 |
| Tybrind Vig 1, Fyn, Denmark | 3310 ± 95 bc | K-3557 |
| Tybrind Vig 2, Fyn, Denmark | 3420 ± 95 bc | K-4149 |
| Maglemosegards Vaenge II, Denmark | 3470 ± 75 bc | K-4336 |
| Maglemosegards Vaenge I, Denmark | 3770 ± 75 bc | K-2722 |
| Korshaven, Mejilo Nord, Denmark | 4310 ± 95 bc | K-2040 |
| Pesse, Netherlands | 6315 ± 275 bc | Gro-486 |

104    Loch Kinellan
128    Lochlea 5
137    Milton Loch

Also worthy of record in this section is the Dumbuck logboat (no. 35) which was found in a nearby 'dock' rather than in the crannog itself. In spite of the controversy over the objects that were supposedly found in it, the antiquity of the boat itself has not been doubted.

Fox (1926, 128) attributes a medieval date to Kilbirnie Loch 1 (no. 75) on the evidence of the bronze vessels that were found in the boat, but this cannot be considered a proven association and must be disregarded as a chronological indicator.

Although the frequently-assumed equation of logboats with crannogs should not be over-stated (Munro 1882, 279-80), the discovery of several logboats built into crannog (sub-) structures justifies an inferred date in the later prehistoric or early historic periods in each case on the assumption that these are redundant boats re-used as beams or infill. Numerous further examples have probably been discarded by excavators as unrecognised or unworthy of detailed record. It should also be borne in mind that the felling of the timber for the boat will inevitably predate (possibly considerably) the construction of the crannog.

The chronology of the construction and use of Scottish crannogs itself remains unclear (Morrison 1985, 22-5; Crone 1993a) and will only be elucidated by extensive further excavation. However, it must be taken as axiomatic that they are long-lived and multi-period constructions. Apart from the evidence from Buston and Loch Glashan that is cited above (nos. 11-12 and A37-44, respectively), Lynn (1986) and Warner (1986) have suggested, on the basis of re-interpretation of the evidence from excavated Irish examples that the earliest structural remains recognised in the crannogs at Lagore, Co. Meath, Ballinderry no. 1, Co. Westmeath and Lough Faughan, Co. Down should be considered as representing phases of occupation rather than construction, as was initially suggested. The earliest phase (1a) of the 'royal' crannog at Lagore is apparently to be historically dated to the mid-seventh century AD and this date is at least broadly supported by the archaeological evidence (Warner 1986, 77). The structural remains of this phase are interpreted by Lynn (1986, 69-71) as comprising the accumulated remains of permanent dwellings (in the form of collapsed wattle-work) within the collapsed remains of a palisade-ring (the 'bow-shaped timbers'), outside which there is an extensive deposit of butchered animal remains. The suggestion is made (Lynn 1986, 72) that the crannog was laid down by dumping during the summer months to a level higher than that of the flooding anticipated over the next winter, but it is unclear how this hypothesis would relate to a crannog of pile- rather than dump-construction. Most of the finds excavated at this site were found in the disturbed upper levels or outside the lines of the palisades. The use of the Lagore crannog in archaeological study as a 'quarry' for dateable artefactual comparanda (Warner 1986) is based around its attribution to either 651 or 677-969 AD on the basis of (possibly erroneous) historical evidence. The Roman pottery recovered is not seen as indicating occupation in the Roman period.

At Lough Faughan (Lynn 1986, 71) the excavator (AEP Collins) noted no sharp distinction between the substructure and occupation layers. The woven brushwood mats laid down under the occupation layers are compared to those found in the substructure at Lagore; these may have been the remains of wicker panels laid down as walkways. Ballinderry no. 1 may be re-interpreted (Lynn 1986, 72-3) as comprising two superimposed crannogs, the early (smaller) structure being subsequently enlarged. Although the number of excavated examples is far too small to form anything approaching a valid sample, it is evident that, in Ireland as in Scotland, the crannog is a type of monument which was long-lived individually and long-lasting as a type. The chronological value of vessels found in association with crannogs must be consequentially reduced.

The recognition of the possible re-use of logboats as troughs in *fulachta fiadh* cooking-places (O'Riordain 1965, 43-5) at Curraghtarsna, Co. Tipperary, Ireland (Buckley 1985; Brindley, Lanting and Mook 1990, 27-8, no. 25a) and Teeronea, Co. Clare (Brindley, Lanting and Mook 1990, 26, no. 2a) may suggest that equivalent discoveries may be anticipated in the (comparable) burnt mounds of Scotland, the chronology of which is discussed under section IV.9.4., below.

### IV.7.3. HIGH ANTIQUITY INDICATED ON GEOLOGICAL GROUNDS

*a. Stratigraphy recorded*

50    Friarton, Perth
?later 6th millennium bc

The date of this discovery must be regarded as suspect as the timber, which was identified as 'Scotch fir', was only recorded by Geikie after discovery and not examined *in situ*.

*b. Noted as found deep in deposit*

28     Dingwall
45     Falkirk
61     Glasgow, Old St. Enoch's Church
64-8   Glasgow, Springfield 1-5
72-3   Glasgow, Yoker 1-2
148    River Carron

*c. Noted as found in high river terrace*

60     Glasgow, London Road     *c.* 9m OD

*IV Synthesis and Analysis*

71   Glasgow, Tontine        *c.* 9.8m OD

Regrettably the exigencies of recording preclude re-assessment and closer dating on the basis of improved stratigraphical knowledge and increased chronological understanding of the drift geology of the areas concerned. The near-universal use of large-scale techniques of mechanical extraction for deep excavations appears to have brought an end to discoveries by this means, but, should any further be made, it should be feasible to date them more accurately than was formerly possible.

*IV.7.4. POLLEN AND VEGETATIONAL DATING*

78   Kilbirnie Loch 4   ?3000-700 bc

It has been suggested that this example probably dates from the Sub-Boreal period on the basis of analysis of the organic remains found in interstices. It must be stressed that this method dates the surrounding sedimentary deposit rather than the boat itself.

*IV.7.5. 'MODERN' FEATURES*

The identification of the following constructional and morphological features may indicate the derivation or skewomorphic copying of specific examples from plank-built boats may be listed as follows:

1. Projecting 'forefoot' or stempost, 'cutwater prow' or 'sharpened stem',

2. False ribs,

3. Inserted ribs, which should not be confused with misinterpreted transverse rows of thickness-gauge holes,

4. False keel,

5. Projecting sternpost,

6. Transom-groove or -board,

7. Washstrakes and extension-pieces.

Toolmarks and repair-patches do not fall into this category of evidence, but the former are considered above (section IV.6).

The following Scottish examples have been recognised:

9 Bowling 1
Projecting forefoot

11 Buston 1
Transom, false ribs, inserted boards, trenails

17 Carse Loch
Transom

19 Castlemilk
Trenails

21 Closeburn
Transom-groove

23 Craigsglen
False rib

25 Dalmarnock
Projecting sternpost

27 Dernaglar Loch
Transom groove, fitted ribs, trenail(s)

29 Dowalton Loch 1
Transom, thole-pins

31 Dowalton Loch 3
Transom-groove

32 Dowalton Loch 4
False rib

36 Eadarloch
False keel

38 Errol 2
Transom-groove

39 Erskine 1
Transom-groove, trenails

51 Garmouth
Transom-groove, trenails, possible fitted ribs

53 Glasgow, Clydehaugh 1
Bow 'snout-like'

54 Glasgow, Clydehaugh 2
Bow as Clydehaugh 1, transom-groove

55 Glasgow, Clydehaugh 3
Bow as Clydehaugh 1

56 Glasgow, Clydehaugh 4
Transom, saw-cut timber

59 Glasgow, Hutchesontown Bridge
Transom-groove, probable false rib

62 Glasgow, Point House
Transom-groove

64  Glasgow, Springfield 1
Transom-groove

65  Glasgow, Springfield 2
Transom-groove, cutwater prow

66  Glasgow, Springfield 3
Cutwater prow

90  Littlehill
Fitted ribs, possible transom-groove

92  Loch Arthur 1
Possible rowlocks, transom-groove

98  Loch Doon 3
Transom, pitch-covering

99  Loch Doon 4
Transom, pitch-covering

100  Loch Doon 5
Transom, pitch-covering

102  Loch Glashan 1
Stempost, false keel, transom

106  Loch Kinord 2
False ribs

107  Loch Kinord 3
False ribs

110  Loch Laggan 2
Probable transom

116  Loch Leven
Transom-groove

127  Lochlea 4
Possible transom

138  Monkshill
Possible stempost, as in a 'coble'

140  'Orkney'
Transom-groove, false ribs

148  River Carron
Stem 'sharp'

149  'River Clyde'
Transom-groove, false ribs.

Particular interest inevitably attaches to those boats that display recognisably 'modern' features and have also been dated by other means:

29  Dowalton Loch 1
Possible extension, transom-groove and thole-pins: found built into crannog.

38  Errol 2
Transom-groove: boat dated by radiocarbon to 485 ± 40 ad.

64  Glasgow, Springfield 1
Transom-groove: boat noted as found deep in deposit.

92  Loch Arthur 1
Possible rowlocks and transom-groove: boat dated by radiocarbon to 101 ± 80 bc.

96  Loch Doon 1
Marks of metal tool: boat dated by radiocarbon to 509 ± 110 ad.

127  Lochlea 4
Possible transom; found in or on crannog.

This evidence suggests that some of the 'modern' features, and particularly the transom-groove, date from at least as early as later prehistory. There is no evidence as to whether this applies to boats with fitted ribs, false keels and coble-like stems, but this view accords well with the evidence of the comparable English radiocarbon-dated examples (McGrail 1978, i, *passim*.):

5  Appleby, Lincs.
Toolmarks (possibly from modern dredging): boat dated by radiocarbon to 1100 ± 80 bc.

11  Barton, Lancs.
Washstrake: boat dated by radiocarbon to 1030 ± 65 ad.

19  Branthwaite, Cumbria
Possible transom or 'backboard': boat (or more probably trough) dated by radiocarbon to 1570 ± 100 bc.

22  Brigg, Lincs.
Transom, 'brackets', 'shelves', 'knees' and 'ridges': boat dated by radiocarbon to 834 ± 100 bc.

55  Hardham 2, Sussex
Ridge: boat dated by radiocarbon to 295 ± 50 ad.

--  Hasholme, Humberside
Beam-ties and side-extension: boat variously radiocarbon-dated to 300 ± 80, 400 ± 90 and 600 ± 100 bc.

70  Kentmere 1, Cumbria
Solid stem: boat dated by radiocarbon to 1300 ± 120 ad.

112  Poole, Dorset

Toolmarks, ridges, false stem and transom slot: boat dated by radiocarbon to 295 ± 50 bc.

141 Walthamstow, Essex
'Bulkhead' and ridge: boat dated by radiocarbon to 1604 ± 54 ad.

146 Warrington 1
'Beak': boat dated by radiocarbon to 1190 ± 60 ad.

147 Warrington 2
Ridges, transverse strengthening timber and thwart or washstrake: boat dated by radiocarbon to 1020 ± 50 ad.

148 Warrington 3
Transverse ridge: boat dated by radiocarbon to 875 ± 60 ad.

149 Warrington 4
Possible thwart: boat dated by radiocarbon to 1072 ± 60 ad.

150 Warrington 5
'Beak': boat dated by radiocarbon to 958 ± 65 ad.

152 Warrington 7
Possible 'beak': boat dated by radiocarbon to 1090 ± 60 ad.

156 Warrington 11
'Beak': boat dated by radiocarbon to 1000 ± 90 ad.

By way of comparison, there may also be noted the discovery of a 'slab of carved wood projecting from the ends of two ancient logboats brought up from the floor of the Baltic Sea between Bornholm and Gotland by Polish trawlers' (Crumlin-Pedersen 1972, 336-7 cited in Hale 1980, 121). The 'fin' of the better-preserved of these boats was 'oblong' in form and measured over 1' (305mm) in length; the sides are 'straight' and a slight downwards inclination was evident below the bottom of the boat. It had been carved from the solid as part of the construction process and notched on the end, possibly to receive a cord running up the endpost. Although the significance of this object is unclear and no date is stated for it, the possibility of its being the transom of a boat of relatively recent date cannot be ruled out.

*IV.7.6. Constructional use of iron or unspecified 'metal'*

The following Scottish examples have been recognised:

6 Barnkirk
Toolmarks

21 Closeburn
Toolmarks

25 Dalmarnock
Holes drilled with probable metal bit

27 Dernaglar Loch
Transom groove, fitted ribs held by trenail(s)

65 Glasgow, Springfield 2
Metal toolmarks

77 Kilbirnie Loch 3
Metal toolmarks

85 Lea Shun
Metal binding strip (may be repair feature)

92 Loch Arthur 1
Toolmarks

94 Loch Chaluim Chille 1
Iron rings

96 Loch Doon 1
Metal toolmarks

98 Loch Doon 3
Possible toolmarks

102 Loch Glashan 1
Metal toolmarks

140 'Orkney'
Toolmarks

*IV.7.7. Dendrochronology*

To date, no Scottish logboat has been dated by the analysis of visible tree-rings, but the building of a tree-ring sequence for South central Scotland from AD 946 to the present (Baillie 1977) may make it possible to date some of the surviving examples by this means.

In conclusion, it is worth repeating that the logboat is a simple form of construction which requires only a low degree technical understanding and a simple toolkit for its construction. It has probably been readily re-invented on numerous occasions when circumstances demanded, and may well be so again. A recently-available seed catalogue (Chiltern Seeds 1986) is probably justified in imaginatively offering Oregon or Red Alder (*Alnus rubra*) as a handsome medium-sized tree which grows fast on the right size and is said to be an 'excellent tree for making dug-out canoes'.

## IV.8 Paddles and Oars

The rare discovery of oars and paddles, and their even more infrequent publication in the archaeological literature (Coles, Heal and Orme 1978, 20, fig. 11 and 34-42, *passim*.) currently preclude a comprehensive analysis of their typology and functions. The compilation of a complete corpus for the British Isles is long overdue, particularly as more examples are revealed by underwater exploration and survey.

The most important recent discovery is that from Canewdon, Essex (Darvill 1987, 53; Wilkinson and Murphy 1995, 152-7) which was found in May 1993 during archaeological reconnaissance of the intertidal zone on the southern shore of the middle Crouch estuary and is now held by the National Maritime Museum. It lay roughly horizontally within what was evidently a saltmarsh deposit of very soft and plastic greyish-brown clay at an altitude of about 2.3m OD. There were no associated artefacts and the paddle was found complete although slightly damaged and divided into three pieces by hairline cracks.

The paddle is of oak and is notable for its length (2.08m), narrow and clearly-defined lanceolate blade (0.63m long by up to 0.14m broad) and (unparalleled) swelling at roughly the midpoint of the shaft. The blade was roughly diamond-shaped in section while the shaft was subcircular, with a maximum diameter of about 50mm. Neither toolmarks, distortion nor evidence of wear were noted, and the standard of workmanship was high.

No close parallels are evident for the paddle which has yielded a radiocarbon determination of 950 ± 70 bc (BM-2339) which may be calibrated to about 1093 BC. It is chiefly remarkable for its unusual length; when held as if to paddle, the swelling on the shaft falls naturally to the lower hand while the upper hand grips the slightly expanded terminal, suggesting its suitability for practical use.

Three methodological problems arise in the study of oars and paddles. The impracticable construction and ornate decoration of some of the paddles from Tybrind Vig, Denmark (noted below) indicate that some, at least, had a non-functional role in ritual, or, more plausibly, as display items. In addition, the similarity of these relatively simple forms to a variety of non-maritime artefacts is a fruitful source of confusion. Chief among these types are mill-paddles, agricultural implements and the structural timbers of crannogs; Scottish examples of all these types are cited. Finally, and possibly most critically, the simple form of these objects renders them extremely difficult of dating when found unassociated or in unstratified contexts, unless radiocarbon dating techniques are used, although erroneous results may still result from the thickness of the timber ('old wood effect') as in the case of the medieval side-rudders from Southwold, Suffolk (Hutchinson 1986). In practice, the frequency of their discovery in areas of deposition or collection of modern maritime debris must make this option appear prohibitively expensive as a general practice.

The problems involved in the propulsion and steering of early and traditional boats have been discussed by McKee (1983, 130-52) and McGrail (1987a, 204-51) who note three means of human propulsion which leave significant artefactual remains:

1. *Poling, punting, quanting or setting* is used in shallow water and involves the use of poles of varying lengths which may have secondary uses as sail-spars or boat-hooks, and may be distinguished by forked wooden or iron ends which prevent their being pushed too deeply into the mud.
2. *Paddling* may be effected from a standing, sitting or kneeling position using either a single- or double-handed action. The restricted beam typical of logboats probably makes this the generally-preferred method of propulsion for this type of craft. Modern ethnographic parallels reveal no apparent pattern of shape or form, and it is perhaps best not to over-analyse so long as only a few dated examples are available. This impression of diversity is reinforced by the varied shapes of the Danish paddles of varied dates that are illustrated by Rieck and Crumlin-Pedersen (1988, 14, 23, 24-5, 40, 126).

Also found in the underwater excavation at Tybrind Vig, Fyn, Denmark (Andersen 1986, 101-5), but not apparently in association with either of the boats, were ten paddles of three different forms. All were one-piece carvings from ash trunks with short heart-shaped blades of different sizes on shafts about 1m long. The blades of two of the larger examples bore incised decoration infilled with an unidentified brown colouring matter. Study of these objects suggests the need to re-assess the objects from Satrup and South Schleswig that have traditionally been identified as spades; each was of ash and comprised a shaft, measuring between 1 and 1.2m in length, and a heart-shaped blade, which was patterned and slightly smaller than is the norm in the case of some, presumably ceremonial, examples.

The most recent English study of paddles in antiquity (Wright 1978, 193-5) covers the two examples that were found at North Ferriby, Yorkshire in estuarine clay and in close proximity to three sewn boats of about 1350 bc. The first paddle was found incomplete in 1939 and destroyed in the fire at Hull Museum two years later after a copy had been made but without an authoritative identification of the timber, which was probably ash. The blade measured 0.85m in length by up to 150mm in breadth and 23mm in thickness, while the diameter of the shaft was 44mm. The damaged blade was of 'elongated spade-shape' with curving, convex shoulders, slightly convex sides, a square-tipped end, and an area of about 0.11 sq. m. The second discovery (in 1946) took the form of a short length from around the junction of the blade and the handle; this object has suffered near-total disintegration, but the timber has been identified as ash. Wright compares the first paddle with those used by the Maori for the propulsion of large logboats and differentiates this from the much larger type used for steering. The thickness of the shaft is seen as inadequate for the greater stresses imposed by use as an oar.

3. *Rowing, sculling or pulling* involves the use of a mechanical lever around and against a rigid fulcrum or pivot which is integrated into the structure of the boat. Although what are probably oarlock or rowlock openings have been identified on the medieval extended logboat from Kentmere, Westmorland, England ((Wilson 1966, 82), the need for considerable working space (in a broad beam) makes this an improbable (if not impossible) form of propulsion for most of the smaller Scottish logboats, and it may be significant that the much larger Hasholme boat was itself paddled (Millett and McGrail 1987, 131-5). The significance of the 'oarlock' noted on the surviving side of the Erskine 6 logboat (no. 44) remains unclear.

In the case of most logboats, propulsion and steering are two aspects of the same problem as the differential application of a propulsive force on one side of the vessel will induce or maintain progress in a desired direction. Exceptionally, however, a specialised steering-oar may be distinguished which is asymmetric in form and is generally mounted in the fore-and-aft plane near the stern with a controlling side-tiller attached. In operation, such an oar induces additional drag on the inside of the turn and thus acts as a differential brake. Even if the mechanical problems of its mounting are disregarded, such an oar would appear appropriate only to a vessel of considerably greater size than any known Scottish logboat, and the Newstead steering-oar (no. A55) must be tentatively attributed to a boat of plank construction.

The major characteristics of the adequately-recorded oars and paddles (including some evidently not intended for boat propulsion) found in Scotland are tabulated as table 9.

## IV.9. *Miscellaneous Discoveries*

In recent years the study of archaeological timberwork has advanced rapidly, most notably through work in the Somerset Levels (Coles, Heal and Orme 1978; Coles 1982; Coles 1984; Coles and Coles 1986; *Somerset Levels Papers passim.*), which has vividly demonstrated the number and variety of wooden artefacts used in British antiquity while Barber (1984) has discussed some of the smaller wooden artefacts that have been frequently discovered (but less often recorded) in Scottish archaeology.

The development of woodworking materials and techniques has been studied in considerable detail by Earwood who summarises (1993, *passim.*) the evidence for (portable) domestic wooden artefacts (excluding, by definition, oars and paddles) with particular reference to their chronology and methods of manufacture. Her analysis is particularly germane to the present study.

She notes (1993, 4, 26) that the recorded distribution of such discoveries (like that of logboats) owes much to the progressive artificial reduction of the formerly-massive waterlogged, alluvial and peaty areas that formerly characterised Scotland down to the relatively small remnants that survive today. Isolated pockets (at least) survive in all parts, and further discoveries may be anticipated within them, while local waterlogging may lead to survival within the pits and ditches of excavated archaeological sites of a variety of types.

As only relatively few such discoveries (43 in all) have entered the archaeological record, their distribution must necessarily be analysed on the basis of relatively large areas (in this case, the post-1975 administrative regions), as in table 10. The figures cited relate to recorded discoveries, rather than individual artefacts, and include a few artefacts which are stave-built rather than of monoxylous construction. They appear, however, to fall into a pattern broadly comparable to that recorded for logboats and characterised by relative infrequency in the arable areas of eastern Scotland, but with a lower proportion in the industrialised central belt and a much higher number in the far north and west, where bog butter containers of various types are most commonly found. There is no available evidence for the deposition of such containers.

On the basis of the available evidence, Earwood sees stave-building or cooperage as possibly developing in Britain and Ireland in the late prehistoric period, but remaining distinctly separate from introduced Roman types. Both turned and carved vessels are commonly found in deposits of this date, but carving was evidently gradually superseded as a method of manufacture (as against decoration) by lathe-turning during the first millennium AD, most vessels being manufactured by turning after about the 6th century AD. During and after this period, carving was, however, necessarily retained for the manufacture of articles of non-circular shape and of either small or large size. Spoons and ladles were apparently commonplace objects, while wooden troughs of various types remained in common use well into the medieval period.

The Scottish evidence for various classes of logboat-related artefact is, accordingly, here summarised, in an attempt to stimulate interest rather than to reach firm conclusions or suggest a formal typology. Indeed, evidence from outside Scotland, suggests that many of these relatively simple timber artefacts were used for purposes other than those originally intended, making a mockery of formal typology.

### IV.9.1. LOG-COFFINS

The close relationship between log- (or monoxylous) coffins and logboats, and the relative frequency with which they have been found to contain archaeological discoveries of the highest importance, render this class of artefact of great interest. For example, the possibilities may be considered that the recorded logboats from Blaenffos,

| Artefact | Length over all | Length of blade | Blade form | Shoulder form | Section of loom, handle or shaft |
|---|---|---|---|---|---|
| A33. Loch Coille-Bharr, paddle | >1.2m | 0.22m | 'like a barbed arrow' | (Unstated) | (Unstated) |
| A36. Loch Glashan, paddle or oar (possible) | Survives incomplete | 0.41m | Straight-sided | Blade tapers into handle | (Lost) |
| A38. Loch Glashan, crannog, paddle | 1.1m | 0.06m | Roughly rectangular with slight midrib | Irregular shoulder on one side, tapered on the other | Roughly circular |
| A39. Loch Glashan, crannog, paddle (possible) | Survives incomplete | 0.09m | Roughly rectangular with flat and convex surfaces | Blade tapers into handle | Spade-type grip |
| A41. Loch Glashan, crannog, model paddle (possible) | 0.21m | 0.12m | Elongated tongue with rounded end and slight keel on one side | Rounded shoulders tapering into handle | Oval, with thickened terminal |
| A46. Loch Kinord, paddle | >0.52m | 0.22m | Roughly square | Concave | Slightly oval |
| A51. Lochlea, crannog, oars (surviving example) | >1.04m | 0.86m | Bilaterally symmetrical | Asymmetric sloping shoulders | Irregularly rounded |
| A52. Lochlea, crannog, double-paddle | 1.37m (sections combined) | 0.6m and 0.42m (latter incomplete) | Roughly oval | Asymmetric sloping shoulders | Roughly oval |
| A54. Newstead, steering-oar | 1.65m | 0.66m | Rounded asymmetric near-rectangle | Tapered | Circular |
| A55. Oakbank, crannog, paddle | >1.35m | 0.7m | Roughly rectangular | Tapered | Rounded |
| A60. Ravenstone Moss, paddles, preserved example | 0.89m | 0.73m | Very broad and highly eccentric with flat section | Pronounced reverse-curved shoulder on one side | Squared with rounded corners |
| A64. Rough Castle, mill-paddle (possible) | 0.34m | 0.23m | Roughly oval | Asymmetric shoulders tapering into blade | Rectangular, transverse to blade |
| A65. Rubh' an Dunain, oar or paddle (possible) | 0.45m | 0.3m | Roughly oval | Diagonal | Rectangular |
| A69. Tentsmuir paddle | 1.2m | 0.53m | Roughly rectangular and curved in section | Asymmetric and concave | Varies, rounded near blade, oval centrally and square near broadened handle |

*Table 9. Summary of sizes and forms of Scottish oars and paddles.*

Pembrokeshire and Chatteris, Cambridgeshire (McGrail 1978, i, 160 no. 13 and 175, no. 29 respectively) should be re-classified as log-coffins, the former having been found within a barrow or natural mound and the latter containing a burial accompanied by a rapier.

It is not suggested that the log-coffin is a type which has an intrinsic cultural or chronological significance, being a simple piece of woodworking without any of the considerations of stability or volumetric efficiency that are necessarily involved in the construction of a true logboat, and which are generally indicated by the presence of thickness-gauge holes, thinner walls and a higher standard of workmanship. The construction and use of such plank-built coffins as those from Culross, Fife (NMRS dataset NS98SE 21), Leith Links, Edinburgh (NMRS dataset NT27NE 44), and Craigmillar, Edinburgh (NMRS dataset NT27SE 123) are not discussed.

Log-coffins were first recognised in the Scottish archaeological literature by Wilson (1851, 461-3) who attributed them to the 'closing period of the Pagan era' on the assumption that they served to replace the stone cists of his Archaic or Bronze period. The chronological implications of the resemblance borne by the examples from Castlehill, Edinburgh (no. A20) to medieval stone coffins was, however, noted.

The subject was further examined by Childe (1935, 106-7; 1940, 129-30; 1946, 117) who saw them as counterparts to stone-built cists and corrected their dating to the Early Bronze Age (and specifically to his Food Vessel period) on the basis of English parallels. While noting their resemblance to the numerous examples of similar date in Nordic and Germanic Europe, he discounts the suggestion of a Nordic intrusion and visualises at least some of the Yorkshire examples as re-used logboats intended to symbolise the maritime status of the 'boat-chiefs of the Food Vessel population'. He also mentions without comment the suggestion by Hawkes that they represent the Egyptian-derived idea of a ritual voyage to the next world.

The number of log-coffins in Scotland is far lower than might be expected given their prevalence in Scandinavia, North Germany and the Netherlands (Coles and Harding 1979. 289, 293, 295, 298, 300, 304-9, 500-1) although this may be due to the relatively ready availability in Scotland of types of easily-split stone suitable for the construction of cists. This is to the detriment of Scottish prehistory as log-coffins are rewarding subjects for scientific dating and typically have a pronounced tanning action which serves to aid the preservation of any organic contents in such environments as the sandy soils of Denmark (Glob 1974, *passim.*). Log-coffins have traditionally comprised an undue proportion of the richer Bronze Age burials on the continent and in southern Britain.

An interesting typological observation is made by Ashbee (1960, 86-91) who discusses the English and Welsh examples, and distinguishes square-ended examples from those that are 'boat-shaped'. He also notes their association with both inhumation and cremation burials, the relative frequencies being about two of the former to each of the latter. Although this distinction must be discounted with the realisation that many proven logboats are square-ended in form, the value of the available chronological evidence and the potential for the indication of the range of forms and uses involved, justifies a summary of the better-known examples from among those English and Welsh log-coffins that are not accepted as possible logboats by McGrail (1978).

Of particular interest are the coffined burials at Loose Howe, Yorkshire that were interpreted by the excavator as boats 'in ritual intention at least' (Elgee and Elgee 1949, *passim.*, esp. 92). They were revealed in 1937 during the excavation of a large and prominent composite barrow in the central part of the North Yorkshire Moors. The two log-coffins lay parallel and close together; neither example was ribbed and there were no thickness-gauge holes but in each case a well-defined external keel joined the rounded bow to the irregular stern. The excavator designated the smaller of the two as a 'canoe'. It measured 2.75m in length by up to 0.68m transversely; the 'starboard' side had retained its bark and the other side was damaged. The larger coffin and its cover were each similar in form to the 'canoe'. The coffin itself was found in a fragmentary condition and measured 2.5m in length by up to 0.63m transversely externally and 1.95m by about 0.23m in depth internally. The lid or cover measured 2.7m by up to 0.78m transversely externally and 2.03m in length by up to 0.2m internally. The coffin and cover together apparently contained space measuring about 1.95m by 0.6m by 0.45m in depth (0.53 $m^3$) within which the body had lain with its head at the WSW end, immediately in front of the 'stern' block. Although there was no direct dating evidence for these coffins, they were found stratigraphically below a secondary cremation which was accompanied by an axe-hammer, a bronze dagger and a pygmy vessel.

Log-coffins may be seen as forming part of the complex of timber mortuary structures that have been studied by Petersen (1970) through the re-analysis of 36 barrow excavations by Greenwell and Mortimer on the Yorkshire Wolds. Petersen notes the existence in certain cemeteries of inhumation burials associated with timber grave-linings, funerary platforms or coffin-like features which are typically associated with Food Vessels, plano-convex flint knives and jet necklaces; no evidence is found for the relative dating of the simple and the more elaborate graves. The burials themselves are typically those of crouched adults and fifteen of the sixteen examples examined were identified as male. Unfortunately the excavators did not distinguish clearly between plank and monoxylous construction but it is unclear whether there is any ritual or chronological significance in this distinction. The Food Vessel has traditionally been dated as a type to approximately 1650-1400 bc (Simpson 1968, 209) and the available radiocarbon dates seem to confirm this chronology.

Brief details of some of the relevant English and Welsh examples (Ashbee 1960, fig. 26) are as follows:

*Beverley, Yorkshire*

In 1846 a log-coffin was found (Gomme 1886, 83) during drainage operations near the town of Beverley and at a point where a 'tumulus' had probably been earlier removed. It measured nearly 8'6' (2.6m) by 4'2' (1.3m) externally, was 7'6' (2.3m) long internally, and is said to have contained 'some fragments of human bones, not calcined'. The lid was formed by a slab attached by four pegs to the bevelled ends of the lower portion.

*Bishop's Waltham, Hampshire*

In 1953-4 the excavation of a bell barrow (Ashbee 1957; Ashbee 1960, 89-90) revealed a roughly-rectangular grave in which had been placed a log-coffin about 5' (1.5m) long, 3' (0.9m) broad and 1' (0.3m) deep. It contained a cremation and two bronze daggers laid upon straw and covered by a sheet of bast; a Food Vessel had also been placed in the coffin. Above the cremation, but within the coffin, there was an anthropomorphic silhouette, which probably suggests the former existence of a double burial.

*Bowthorpe, Norfolk*

In 1979 the excavation of a ring-ditch at Bowthorpe, Norwich (Lawson 1986) revealed a barrow surrounded by two concentric ditches. The barrow covered a central coffined burial, besides ten satellite burials, there was found a single grave beneath the outer ditch.

The central grave was oriented SE-NW and contained the indistinct remains of a contracted inhumation; the only finds were a few flint flakes and 'pot-boilers'. The soil-stain of the central coffin has been interpreted as being compatible with a hollowed trunk split longitudinally and furnished with a 'pillow'. The form is roughly that of a logboat, one end being square and the other pointed.

*Cartington, Northumberland*

Before 1813 a log-coffin (which is now in Bamburgh Castle Museum) and three fragments of its 'cover' or lid were found during ploughing on 'the summit of an eminence' at a height of about 200m OD (*PSAN*, 3rd series, vi (1913-14), 35, 79-84). It was aligned E-W within a 'grave' which measured 4' (1.2m) in depth and had presumably been surmounted by a barrow or cairn.

The coffin was worked from a single oak trunk which had been split to form both the coffin and its lid; distinct toolmarks were noted. It measured 5'5' (1.7m) externally and 4' (1.2m) internally in length, by between 1'8' (0.5m) and 2'1' (0.6m) in breadth 'from edge to edge'. There was a 'dais or pillow' at the E end and the hollowing of a 'much wider' space for the 'back and knees' probably suggests that it was intended to contain a crouched inhumation.

The position of the few teeth found suggested that the body had lain on its left side with the head to the E.

Within the coffin there were noted two layers of clay, the lower of them 'unctuous' and the upper rich in charcoal. Within one of these layers there were distinguished a 'distinct cast' of a leg bone, fragments of kid or calf skin, and a finely-flaked flint scraper. There were fronds of bracken in the lower clay layer and near the W end, at the presumed location of the feet, there were three fragments of a 'Drinking Cup', which was possibly an AOC Beaker.

Radiocarbon assay (Jobey 1984) of a sample from the outer growth rings of the coffin has yielded a date of 1840 ± 65 bc (GU-1648), which determination may be calibrated to about 2220 BC.

*Disgwylfa Fawr, Dyfed*

In 1937 two 'dug-outs' were revealed by the excavation of a complex barrow on Plynlimon (Forde 1939; Green 1987). The larger was of oak and measured about 2.5m in length; it was probably sited on the old ground surface and possibly originally contained an extended inhumation which had disintegrated in the acid conditions. The smaller example was found at a higher level in the barrow and measured just over 1m in length; on it there were found a cremation, a flint blade and a Food Vessel of Yorkshire Vase type.

The two log-coffins have yielded radiocarbon dates of 1910 ± 70 bc (HAR-2187) and 1350 ± 80 bc (HAR-2677) respectively.

*Gristhorpe, Scarborough, Yorkshire*

In 1834 a composite barrow of clay and stone was excavated to reveal a log-coffin beneath a layer of oak branches (Gomme 1886, 81-3, 154-61; Ashbee 1960, 86, 88). The coffin measured about 7' (2.1m) by 3' (0.9m) and was 'formed of a portion of the rough trunk of an oak tree' which had been 'hewn roughly at the extremities, split, and then hollowed internally to receive the body'. The trunk had been bisected to make the coffin and its cover, which was 'of nearly the same dimensions'. The cover was not fastened to the coffin and into the bark at one end there had been cut 'the rude figure of a human face'. The small hole that was situated near the centre of the bottom had presumably served to gauge the thickness of the timber during construction and, possibly, also for drainage.

Within the coffin there was a flexed inhumation which was accompanied by a bronze dagger, three flint arrowheads, three horn ornaments or pins, a wooden pin, a wicker basket and a quantity of unidentified vegetable matter.

*Hove, Sussex*

In 1857 the destruction of a barrow during construction

|  | Finds from crannogs | Finds excavated from other archaeolog-ical contexts | Bog butter related discoveries | Other stray finds | Total |
|---|---|---|---|---|---|
| Shetland Islands Area (2%) |  |  | 1 |  | 1 |
| Orkney Islands Area (5%) |  |  |  | 2 | 2 |
| Western Islands Area (12%) |  |  |  | 5 | 5 |
| Highland Region (28%) |  |  | 4 | 8 | 12 |
| Grampian Region (2%) |  |  |  | 1 | 1 |
| Tayside Region (5%) | 1 | 1 |  |  | 2 |
| Strathclyde Region (28%) | 4 | 1 | 1 | 6 | 12 |
| Central Region (2%) |  | 1 |  |  | 1 |
| Fife Region (2%) |  | 1 |  |  | 1 |
| Lothian Region (Nil) |  |  |  |  | Nil |
| Borders Region (5%) |  | 2 |  |  | 2 |
| Dumfries and Galloway Region (9%) | 2 |  |  | 2 | 4 |
| **Total** | 7 (16%) | 6 (14%) | 6 (14%) | 24 (56%) | 43 |

*Table 10. Summary of circumstances of discovery of domestic wooden vessels and similar artefacts in Scotland.*

work revealed a log-coffin which measured '6 or 7 feet' (1.8 - 2.1m) in length, had been worked from oak and 'shaped with an axe' and lay E-W (Curwen 1954, 152-4; Ashbee 1960, 89; Renfrew 1973, 222).

In it there were the remains of what was probably an inhumation accompanied by an amber cup, a polished stone axe, a possible whetstone and a bronze dagger of Camerton-Snowshill type. Radiocarbon assay (*Radiocarbon*, 18 (1976), 27) of the coffin has yielded a determination of 1239 ± 46 bc (BM-682) which may be calibrated to between about 1475 and 1454 cal BC.

*Newbarn Down, Isle of Wight*

In 1978 excavation of a round barrow within the Gallibury Heap cemetery revealed a complex series of mortuary structures (Tomalin 1979). Among those attributed to the primary phase, there was a log-coffin which had been placed in a D-shaped pit. It measured 2.5m in length and contained a flint knife, several scrapers and a Food Vessel.

*Quernmore, Lancashire*

In 1973 what were probably the unstratified remains of a shroud-wrapped log-coffin burial of Early Bronze Age date were revealed during construction-trenching in the Forest of Bowland (Edwards 1973; McGrail 1978, i, 93).

The two log-coffins that were found may have been joined to form a single example. The more complete of them was rounded at one end and square-cut at the other, measured 2.4m by 0.4m and was set into a depression. The other was damaged but amenable to reconstruction. Two pieces of cloth found were opened to reveal keratinaceous remains.

*IV.9.2. TROUGHS, KEGS AND BOG BUTTER CONTAINERS*

Wooden containers of any sort have rarely entered the British archaeological record, and then only under conditions unusually favourable for preservation. A notable example was the damaged 'tub' containing a founder's hoard of socketed axes and other objects that was found in 1939 at Stuntney in the English fenland (Clark and Godwin 1940), when it was believed to be the only wooden domestic vessel of pre-Iron Age date from England or Wales. The body and groove-retained base were worked from alder and measured about 12' (0.3m) in diameter. Clark compared this object both with Irish bog butter kegs and with the various vessels from the Glastonbury lake-village.

In an interesting demonstration of the dangers of classification unsupported by chronology, Clark attempted to derive a typology of wooden vessels from the apparent parallels with the development of boats, by equating simple logboats with single-piece domestic vessels, logboats with groove-mounted transoms with two-piece vessels 'with separate base', and plank boats with stave-built vessels. The two-piece vessel is seen as an intermediate form in the transition to the lighter and more easily-handled stave-built type, the existence of which was suggested by the form of the decoration on some of the handled Beakers from the area.

Among the characteristic artefacts of Scottish ethnography are the troughs, querns and kegs in which butter, tallow or similar substances were placed, presumably (but by no means certainly) for cool and long-term storage in peat (Fenton 1976, 150). That such butter left unrecovered was in time converted to adipocere or 'bog butter' was revealed by the early use of scientific analysis (Macadam 1882). Containment in wooden troughs, querns or kegs was not necessary for such storage as wrapping in cloth, bark, rushes or, possibly, skin proved adequate in many, if not most, cases. As the likelihood of such remains being recognised in peat-digging is low, it is probable that many such deposits have been destroyed without record. The current state of the study of this material has most recently been summarised by Earwood (1993, 12-15) who notes that there is archaeological evidence for such bog-deposition in Scotland and Ireland while related practices are attested ethnographically in Wales and the Faeroes. The containers so used vary widely in capacity but may be divided between cylindrical kegs (of carved, turned or stave-built manufacture), monoxylous troughs and a variety of smaller bowls and dishes which were presumably not specifically intended for burial. Most, if not all, of the recorded containers are of flat-bottomed form while features suggesting intentional manufacture might include perforated lugs (for lowering the container into the bog) and the presence of a lid or string-groove (for retaining a fabric covering). Wicker containers and leather or fabric wrappings have also been recorded.

Earwood (1993, 14-15) notes the relative infrequency of such discoveries, the 36 bog butter containers recorded from Scotland and Ireland representing only about 20% of the total number of containers recorded from those two countries. The eight recorded Scottish finds are all from the more rugged areas of the far north and north-west, the largest group (three) being from Skye.

The major characteristics of the best-recorded such discoveries are cited in table 11 and may be compared (table 12) against those from the broadly-comparable settlement or 'lake village' at Glastonbury in the Somerset Levels (Coles and Minnitt 1995, and refs.) which was identified in 1892, when its surface remains comprised about 60 mounds spread over an area of about 1ha. It was early recognised as a site of exceptional importance on account of its size, good preservation conditions (for both artefactual and faunal evidence) and wetland location. Subsequent excavations and environmental investigation by Bulleid, St George Gray and others were chiefly concerned to reveal timber features, clay floors and hearths on a massive timber substructure. The excavators typically paid greater attention to artefacts than to structural timbers, but there were deficiencies in the study of even these. Publication was extensive and considerable numbers of finds recovered were subjected to advanced scientific analysis and selected examples have survived, but the conservation techniques used were of uncertain efficacy and the standard of recording was consistently poor; it is uncertain what proportion survives of any class of artefact and stratigraphic relationships were not recorded.

This relatively large but ill-recorded body of material has been subjected to repeated re-interpretation in the light of successive waves of archaeological theory. Most recently, Coles and Minnitt have used the excavation records, small-scale excavation and supplementary scientific investigations (including radiocarbon assay and pollen analysis) as the basis for a comprehensive and detailed re-interpretation.

The village was early recognised as an essentially-peaceful permanent settlement of round timber houses built on an artificial timber structure within a defining feature. An area of 8900m$^2$ was excavated and some 1000t of deposits removed in excavations which were frequently hampered by flooding, so that investigation of the substructure was presumably not extensive. No attempt was made distinguish phases of occupation or to identify patterns of seasonal activity or variations in structural form. The 90 individual mounds that formed the basis for the site were built on layers or spreads of brushwood and logs (mainly alder) which were laid criss-cross and supplemented by dumps of rush, bracken, peat, rubble and (imported) clay to form a cushion rather than a framework. Some of the logs were dressed or had mortice-holes, and few artefacts were found in this layer. The typical house was round on plan and had a wall of stakes driven through clay into the substructure; these stakes were than linked by woven rods or panels and daubed before the house was roofed with rush, heather or straw. Slab-hearths were identified and there was some evidence of timber flooring. Further short curving lines of stakes and posts were seen as indicating windbreaks or shelters while areas of spread stone and clay were recognised between the houses; these were presumably replaced repeatedly. Bulleid saw the village as being defined by a surrounding feature of alder posts, which is termed 'fencing' by Coles and Minnitt; on the E there was a 'causeway' which incorporated a possible landing-stage.

Bulleid emphasised the watery context of the settlement, which he saw as being located on a relatively firm area of peat in running water, while the deposits outside the village were of a rushy character; in essence, it was a true wetland site set in swamp, rather than a flooded dry settlement. As such, it exploited a wide range of resources within a complex 'concave landscape'. and rafts or logboats must have been essential for transport.

Subsequent investigations have defined the following stratigraphic sequence:

(7) 'causeway' and 'embankment' outside (6),

(6) timber palisade or 'fence' outside (5),

(5) houses, on (4),

(4) layers and floors of clay,

(3) timber substructure,

(2) alder/willow fen carr,

(1) (basal) *Phragmites* reed peat.

Outside the settlement, the (alder) fen carr was found to have had brackish and detrital tendencies. It was succeeded by a fine detritus mud resulting from a wetter climate at about 2800 BP and reaching a climax three centuries later, and which represented the extreme southern penetration of Godwin's 'Romano-British' estuarine clay. This was succeeded, after 1660 BP, by the development of a post-peat environment of alluvial silty clay. The depositional regime was organic rather than mineralogenic, while the presence of perch as the major fish species implies relatively clear water.

Re-interpretation by Coles and Minnitt has allowed the recognition of constant change and evolution through the processes of repair, rebuilding and abandonment in the life of the settlement, demonstrable changes in use or shape being recognised in 35 mounds. They define four radiocarbon-dated phases which are intended to be seen as snapshots rather than clearly-defined periods of occupation:

'Early'
This phase is seen as comprising six houses, fifteen spreads of clay and 42 floors in all, two of the houses being replacements for others destroyed by fire. It was built within the fen carr and apparently comprised four units, each of one house with adjacent clay spreads and adjacent storehouses or racks. Only limited activity and limited specialisation can be discerned within this phase which probably had a population of about 50 and a lifespan of about one generation in about 250-225 BC.

'Middle'
This phase is seen as one of expansion, a further four or five families joining to village to increase the population to about 125 and to occupy eleven houses and thirty clay spreads with 58 separate floors, over a period spanning two generations between about 225 and 175 BC. No special areas for communal activity were identified in this phase which coincided with volcanic activity which may have precipitated significant environmental change.

'Late'
The settlement continued to expand throughout this phase which lasted for about three generations between about 175 and 75 BC, when it was occupied by about fourteen self-sustaining units, each comprising a house and attendant yard with large shelters or fenced spreads attached. The total number of houses rose to thirteen with six sheltered areas, 57 clay spreads and 100 clay floors in all, and the population was probably about 200. The 'fence' was complete by this stage, enclosing an additional area on the E, where the embankment was turned into the 'causeway'. This large settlement must have had a major impact on the local environment and resources.

The domestic and industrial activities of the settlement were at their most varied in this phase, to which is attributed the bulk of the pottery and most of the wooden artefacts. One roundhouse (M9) of the middle phase was apparently converted in this period into an elongated structure which may have formed a social focal point. There is no evidence of great wealth but a small bronze hoard and a bronze bowl are attributed to this phase. Midden deposits were laid down outside the entrances and burials were deposited outside the fences (adults) and within the settlement (neonates).

'Final'
This phase is characterised by initially gradual and subsequently increasing contraction. Only five houses (all of them renewals) were built and spreads of clay laid down while the longhouse M9 was re-adapted to the round form. The causeway was possibly abandoned and the level of domestic and industrial activity declined; only one quern is attributed to this period and metalworking is scarcely represented. Seasonal occupation and the exploitation of purely local resources by a population of about 50 is suggested in the period 80-50 BC. This process may be the result of fundamental palaeoenvironmental change resulting in a slow rise in water levels in the Upper Brue valley, with increased valley run-off and resultant deposition through silting damaging the economic basis for the settlement rather than its physical condition.

The vast bulk of the material assemblage was domestic in nature. Finds of possible weapons or military equipment were extremely rare and the many pieces of clay slingshot found were probably intended for hunting or fowling. Several tonnes of pottery (comprising over 5200 vessels) were recovered, but no kilns or wasters and only 27 fragments of glass or vitreous material. Shale may have been worked on the site, flint was of little importance and the stone assemblage comprised mainly quernstones, whetstones and spindle whorls.

There is evidence for a modest level of both ferrous and non-ferrous metalworking, in both cases probably smithing rather than primary smelting. Many crucible fragments and smithing fragments were found, but little slag and no mould fragments or metalworking tools (except files and whetstones). Lead and tin objects recovered included sinkers and two iron currency-bars may have served as raw material for smithing.

Bone and antler were worked in considerable quantities, using saw, knife, drill, gouge and chisel as well as rasp and file. There was prolific evidence (particularly bobbins, spindle whorls and loom-weights) for all stages of textile manufacture.

Study of the timberwork and wooden objects reveals that the 'fence' or revetment was of post-and-rail type with mortised joints and that some of the internal foundation timbers incorporated double or triple holes containing vertical poles. Some beams were found re-used as foundation materials, and had probably been pegged down. Portable objects discovered included a ladder, lap-jointed wooden timbers and non-structural pieces of high quality. Few tools and little direct evidence of their use survive. The marks of axe or adze, gouge or chisel and auger have been recognised as have those of the saw on wood of diameter 150mm or less; it is assumed that larger pieces were split rather than sawn. There is also evidence of charring. Portable wooden artefacts were surprisingly few in number, although this is possibly owing to discarded objects having been burnt as fuel rather than discarded into pits as would be the practice on most sites. Most of the objects that survived to excavation were found outside the fence or were incorporated into floor foundations. Only a small proportion evidently survived of a wide range of types, which were mainly domestic. Handles and containers were the most common discoveries, incorporating stave-built, lathe-turned, hand-carved and bentwood examples which were made with care and often decorated. Oak (followed by ash) was the most commonly identified timber for the manufacture of portable objects by contrast with the timber used for foundations which was primarily alder, followed by oak, ash, willow and birch. Woodworking tools discovered included seven adzes, three gouges, four saws and single examples of a spoon augur, a wooden mallet and a draw knife; twelve iron knives were also found.

Following the publication of a comprehensive account (Earwood 1990) a descriptive and quantitative summary of the recorded examples of oars, paddles and monoxylous troughs and other artefacts germane to this study is given (as table 12) to indicate both the variety of forms and techniques, and the relative frequency of their manufacture by monoxylous reduction.

The major impression gained from the study of this deposit is the wide range of forms of small monoxylous vessel found in it, and (by inference) the wide range of uses to which they were put. By way of comparison, Earwood points out that there is a complete lack of stave-built containers in excavated Irish sites of this period, as at Loch Glashan.

*IV.9.3. MILL-PADDLES AND TROUGHS*

One of the distinctive building-types of Scottish vernacular architecture is the 'horizontal' watermill, in which the mill-paddles or 'feathers' rotate in the horizontal plane around the vertical axis formed by the axle or 'birl', and drive the upper stone directly, without benefit of gearing. The usual operation of the hopper by contact with a wooden tongue on this upper stone is the origin of the traditional name of 'click' or 'clack' mill for this type (Fenton 1976, 102-3, 105; Fenton and Walker 1981, 34-5, fig. 154). Typically the mill is set aside from its stream and the speed of rotation is controlled indirectly by the operation of the sluice.

The origin of the type is obscure but Irish examples have been noted from as early as the 7th century AD (Barry 1988, 26); the alternative designation as a 'Norse' mill is a misnomer. Examples have been used into recent times along the W coast of Scotland from Galloway to Shetland, a distribution which probably reflects their suitability for the exploitation of small streams, but they are most commonly associated with the Far North, where they have been studied in greatest detail (Fenton 1978, 396-410).

Two items of equipment characteristic of this type of mill require consideration. The more distinctive is the mill-paddle itself which was set in one or, less commonly, two 'flights', each typically comprising about twelve in number. Unlike the paddles of a vertical mill-wheel, they were unsupported by external framing. The angle of set apparently depended on the tradition of the area rather than upon any understanding of hydrodynamics. Such paddles were prone to damage by stones and debris and were fashioned for easy replacement. The low speed of rotation would make out-of-balance loads of little significance.

Study of the examples from Bankhead (no. A7) and Gutcher (no. A26) in the Royal Museum of Scotland (Maxwell 1956) serves to make the point that there is little risk of confusion between this type of paddle and those used for waterborne propulsion or steering. The preserved examples are both markedly dished and scooped, and have been fitted with retaining holes and collars instead of a handle. They are also noticeably larger, particularly in breadth.

Another artefact characteristically associated with horizontal mills is the trough or 'troch' that was set in a sloping attitude between the sluice and the paddles. Some examples narrowed at the point where the flow debouched so as to speed the water flow and increase the pressure; such artefacts were of necessity soundly constructed, and it is presumably for this reason that they were frequently hollowed out as open-ended troughs, rather than built up

|  | Length external (internal) | Breadth external (internal) | Depth external (internal) | Capacity | Remarks |
|---|---|---|---|---|---|
| A6. Baile-meonach, Mull, trough | 1.3m | 0.48m | 0.5m | *c.* 300 lit | Bog butter (possible) |
| A10. Cairn-side, mill-trough | 2.9m | 0.6m (0.4m) | 0.4m | *c.* 450 lit | Mill lade of squared section |
| A11. Cnoc Leathann, Durness, trough | 0.73m (0.48m) | 0.29m (0.19m) | 0.11m | *c.* 10 lit | Bog butter, squared section |
| A13. Cunnister, Yell, trough | 0.5m | 0.3m | 0.18m | ?*c.* 0.25 lit | Bog butter trough, oblong |
| A18. Eadarloch, trough | 1.67m (1.39m) | 0.3m (0.25m) | 0.26m (0.16m) | *c.* 55 lit | Bog butter, squared cavity |
| A24. Gleann Geal, keg | 0.6m | 0.4m (diameter) |  | *c.* 75 lit | Bog butter, cylindrical |
| A28. Kilmaluag, Skye, keg | 0.5m | 1.2m (max. diameter) |  | *c.* 550 lit | Bog butter, cylindrical |
| A30. Kyleakin, Skye, keg | 0.4m | 0.3m (max. diameter) |  | *c.* 28 lit | Bog butter; barrel |
| A41. Loch Glashan, crannog, trough 1 | 0.94m (0.78m) | 0.28m (0.21m) | 0.14m (0.12m) | *c.* 19 lit | Rectangu-lar, flat-bottomed |
| A42. Loch Glashan, crannog, trough 2 | 0.43m (0.34m) | 0.23m (0.2m) | 0.09m (0.07m) | *c.* 5 lit | Rectangu-lar, flat-bottomed |
| A43. Loch Glashan, crannog, trough 3 | 0.7m (0.5m) | 0.29m (0.15m) | 0.13m (0.12m) | *c.* 10 lit | Pear-shaped, flat-bottomed |
| A53. Lochlea, crannog, trough | 1.0m | 0.6m (0.4m) | 0.3m | *c.* 70 lit | Rounded, flat-bottomed |
| A54. Midtown, trough | 0.6m (0.5m) | 0.4m (0.3m) | 0.3m | *c.* 15 lit | Bog butter, 'oblong' |
| A60. Plockton, keg | 0.46m | 0.28m (diameter) |  | *c.* 28 lit | Bog butter, cylindrical |
| A70. Torr Righ Mor, Arran, trough | 1.4m | 0.5m | ?*c.* 0.2m | ?*c.* 80 lit | Bog butter (possible) |

*Table 11. Summary of the forms, dimensions and capacities of bog butter troughs and kegs, the mill lade from Cairnside and selected troughs, bowls and dishes of probable domestic origin being included for comparison. All dimensions are cited in metres, except for the capacity which is in litres. Internal measurements (where available) are cited beneath the relevant external figure.*

from planks in the fashion normal for the lades of vertical-wheeled mills.

### IV.9.4. COOKING- AND BOILING-TROUGHS

The ancient cooking-places of Scotland are recognised and characterised by the substantial heaps of fire-cracked stones and carbonaceous ash that have earned them the traditional designation of *burnt mounds*; they appear generally similar to the numerous *fulachta fiadh* of Ireland (O'Riordain 1965, 43-5; O'Kelly 1989, 223-7). Typically, they take the form of a horseshoe-shaped stone mound of considerable size, with the open end facing a stream or other source of non-saline water. The mound covers a hearth of flat stones and a cooking-trough formed of clay-luted slabs, clay-luted hazel posts, planks or a dugout trough. There appear to be no grounds for attributing any typological or chronological significance to the type of trough used, and the rock most commonly used is sandstone. The traditional, and still generally-accepted, theory is that these structures were used for aceramic cookery using stones heated on the hearth as pot-boilers to heat water in the trough and the mound was thus formed piecemeal from discarded and shattered stones.

That such a practice survived into comparatively recent times is evidenced by Burt (Jamieson 1876, ii, 271-2) who notes that in the Western Isles in the later eighteenth century the 'meaner sort of people' used to 'put water into a block of wood made hollow by the help of the dirk and burning; and then with pretty large stones heated red-hot and successively quenched in that vessel, they keep the water boiling, till they have dressed their food'.

Recent study (Barfield and Hodder 1987; O'Drisceoil 1988) has re-considered the theory that burnt mounds are the remains of structures built for steam-bathing and fumigation rather than cookery. Excavation (in 1980) of two examples in Birmingham revealed a total absence of animal bone or artefacts associated with cooking or settlement in the Cob Lane mound which has been dated to 1190 ± 90 bc. Although these examples are situated at a considerable distance from the main concentration, future field survey in Northumbria and Northern England may reveal further examples in a manner comparable to the numerous additional examples revealed by recent fieldwork in southern Scotland (NMRS *passim.*).

As regards the Irish examples, the most recent summary retains the traditional view of their use, accepting the literary evidence that they were primarily used for cooking although a multi-functional explanation involving secondary use for bathing and industrial purposes cannot be ruled out. Practical experiment (Hedges 1976, 72-3; O'Kelly 1954; Coles 1973, 52-4) has demonstrated the efficiency of this method of boiling water while field survey has demonstrated their frequency of occurrence in every county except Dublin. Radiocarbon evidence has indicated a chronological range between 1900 ± 30 and 510 ± 50 bc (O'Drisceoil 1988, 672), although these results may be considered incomplete, given the small sample.

Most significant is the recovery of environmental evidence from the excavation (in 1981-2) of the burnt mound at Fahee South in the limestone burren country of County Clare (O'Drisceoil 1988, 675-7) which has been dated to 1100 ± 35 bc. Although the preservation of faunal material was poor, enough survived to indicate the consumption of butchered young deer and bovids from domestic stock, and hence their use in conjunction with at least semi-permanent sites rather than as the short-lived 'deer roasts' of hunting parties.

In Scotland, this type of monument has generally been considered of little antiquarian interest and has been susceptible to easy destruction, generally by the removal of material for land infill. Frequently they have been removed without recognition by field surveyors, and the recent discovery of at least eighty examples in the East Rhins of Scotland (RCAHMS 1987, 49-56, nos. 230-303) together with numerous others in Eastern Dumfriesshire and elsewhere (NMRS *passim.*) must indicate that they were formerly more widespread across the country (if not nearly universal) invalidating the traditionally-recorded concentration in Shetland, Orkney and Caithness. Taking into account the recorded occurrences in Western England and South Wales, it is probable that they once existed wherever suitable sandstone was to be found in the highland zone of Britain.

Where they survive, they do so in considerable numbers, some 230 (or one for every 4.5 km$^2$) being recorded in Orkney and at least 200 in Shetland. This density may be expressed as one to every four or six square kilometres, including marginal land, and must be one of the densest-known concentrations of a specific class of monument.

Two peat-fuelled examples have been excavated (Hedges 1975) at Liddle I, Isbister and Beaquoy, Dounby (NMRS datasets ND48SE 2 and HY32SW 11, respectively) in the now-treeless Orcadian landscape. That at Liddle was of exceptionally large size but is of little relevance to this study as the central trough and gully were of flagstone construction; no timber artefacts were found. At Beaquoy, the mound, which was considerably smaller, was found to cover two buildings, which had probably been occupied in succession. No evidence for a trough was found in building I, suggesting the former presence of a timber example, while in building II there was found evidence for a clay-lined sunken trough.

Liddle I has yielded radiocarbon determinations of 876 ± 75 bc (SRR-701) and 958 ± 75 bc (SRR-525) while Beaquoy has given figures of 511 ± 80 and 1677 ± 65 bc (SRR-599 and -1001 respectively) by the same method (Renfrew 1985, 270), while a programme of thermoluminescence assay for five Orcadian mounds (Hedges 1975, 82-4) has indicated a range between 1000 and 400 BC, placing them in the Middle and Late Bronze Age and the pre-broch Iron Age.

In 1987 a further seven burnt mounds were investigated by the Scottish Central Excavation Unit in the East Rhins of Wigtownshire. That at Dervaird, Glenluce (NMRS dataset NX25NW 46) measured about 16m by 8m over all, and was situated in a stream-meander some 10m from the remains of a second mound. Excavation between the two enclosing arms (Russell-White and Barber 1987, 58-60) revealed a cooking-pit cut into the underlying clay, the sides of which were partially revetted with discontinuous stone slabs. The base of the pit was formed by a large and well-preserved oak timber of plano-convex section, which measured about 2.25m in length by up to 0.75m transversely, and between 10 and 17mm in thickness; the flat surface lay uppermost. This timber, like the stone slabs, was probably intended to revet the pit rather than to provide a watertight lining.

In his summary of the general characteristics of burnt mounds, Hedges (1976, 63) notes that the trough is the feature most commonly recognised. It typically measures between 0.7 and 1.2 cubic metres in volume, which equates to a capacity of about 100 litres. The type of construction varies with local traditions and the availability of materials, but the following have been suggested at specific sites:

1. unlined pit dug into impermeable or waterlogged

|  | Loch Glashan | Glastonbury |
|---|---|---|
| Containers (including troughs, bowls and dishes) - total | 7 (1%) | 12 (13%) |
| Monoxylous containers | 5 (5%) | 5 (6%) |
| Stave-built containers | 1 (1%) | 3 (3%) |
| Turned containers | 1 (1%) | 4 (4%) |
| Container lids and bases | Nil | 1 (1%) |
| Bentwood boxes | Nil | 3 (3%) |
| Structural fragments (including dowels, stakes, pegs and pins) | 43 (44%) | 26 (29%) |
| Spoons and ladles | Nil | 2 (2%) |
| Handles | 27 (28%) | 12 (13%) |
| Paddles (including objects of similar form) | 2 (2%) | Nil |
| Boats and related objects | Nil | 2 (2%) |
| Wheeled vehicles (fragments including hubs, spokes and axles) | Nil | 7 (8%) |
| Ard-shaft | Nil | 1 (1%) |
| Miscellaneous (including tools) | 19 (19%) | 24 (27%) |
| Total | 98 | 90 |

*Table 12. Comparative summary of the numbers of artefacts of various type from the Loch Glashan crannog and from the Iron Age lake village aat Glastonbury, Somerset (after Earwood 1988).*

ground,

2. as (1) but lined with a non-watertight lining of stone or wood, the latter being of either plank or log construction,

3. dug-out or monoxylous troughs,

4. watertight lining of animal skin, or

5. bronze cauldron set in a shallow depression.

There appears to be little reason for any technical sophistication in the construction of monoxylous cooking-troughs. Neither trim nor balance would affect the utility of the object, and thick sides would serve a useful function as insulation. Thickness-gauge holes, while not necessarily detrimental, would appear unnecessary, and there is no reason for the ends not to be left complete. No internal fittings would be needed, and the interior could be left rough. This crudity of construction will probably serve to differentiate such discoveries from the more sophisticated design and better workmanship typical of the logboat.

Particular interest attaches to the discovery of what was apparently a logboat re-used as cooking-trough during the excavation of one of a group of seven burnt mounds at Curraghtarsna, Co. Tipperary, Ireland (Buckley 1985). The trunk had been split down its length to form the base and two sides before end-plates formed from unworked tree-trunks were pressed into place. During its use as a cooking-trough part of one side collapsed inwards and was replaced by clay-luted stones. In its original form the capacity of the trough was about 950 litres but the repair would have reduced this figure by over half. A probable parallel from Branthwaite, Cumbria is discussed under section IV.9.6, below.

### IV.9.5. SALT-MAKING TROUGHS

No proven timber artefactual relics of the extensive Scottish coastal salt-making industry have yet entered the archaeological literature, probably on account of the distinctive working practices in the unsophisticated Scottish industry which traditionally produced a low output of a relatively low quality product without the use of such equipment (Whatley 1987, 3-4, 9-10). Rock- or stone-cut 'bucket-pots' were used for supply storage, wooden or iron 'pan spouts' as pipes, a wooden or iron 'pond' cistern for further storage and, generally, an iron 'pan' for the boiling itself. However, the publication (Sale 1981; McNeil 1983) of the remains of open-pan evaporation equipment from Nantwich, Cheshire serves to indicate another type of trough-shaped wooden object with which a logboat might be confused.

The excavation (under conditions of good organic preservation) of two medieval 'wich' (salt-making) houses during urban excavation (in 1979-80) revealed a variety of items of wooden industrial equipment, stake-and-wattle walls and much hazel-work. Although the lead pans had been removed, three brine-boiling hearths were identified in association with a clay-puddled cistern which was replaced (in the later phases of house 2) by two timber troughs or 'ships'. These were evidently intended for the storage of incoming brine and are differentiated in documents from clay-lined troughs ('kinches') and clay-puddled troughs ('troughs').

The better-preserved of these objects (excavation find number W5) was of oak and measured at least 8.3m in length over all; the maximum internal and external diameters were 0.75 and 0.95m respectively, and the depth was 0.4m at greatest. The W end was cut off straight while the E end had probably been blocked or closed by a gate, but this end had been damaged by decay and trenching. The 'ship' had been sunk into the clay up to the top of the sides and the clay had been patched during its long period of use, probably in the 15th or 16th centuries. The fill of the 'ship' was found to comprise rake-out from the hearths, and, at the time of its abandonment, it had been divided into unequally-sized compartments by wooden cross-spars.

*IV.9.6. RETTING AND INDUSTRIAL TROUGHS*

There are no recorded discoveries of retting or industrial troughs in Scotland, but one Cumbrian discovery has been tentatively identified as such, and forms a valuable comparison.

In 1956 a wooden object of trough form was found (Burns 1972) during drain-clearing operations on the edge of a former lake at Branthwaite near Workington. It was re-buried until 1971 when it was further investigated and found to measure 7' (2.1m) in length by up to 2' (0.6m) in beam; at least some of the bark remained in place. A radiocarbon sample taken at this time yielded a date (Q-288) which has been differently cited on several occasions, but appears in the standard account as 1043 ± 110 bc (Ward 1974, 25). The object was tentatively identified as a split and damaged section of 'a hollowed canoe with a carefully in set back board' (Ward 1974, 21) which had been converted into a trough by cutting off one end in such a way as to leave a tenon 8' (203mm) long projecting from the base; a D-shaped hole had been cut through the tenon and a groove had been cut at one end of the trough to retain a carefully-shaped board. The acceptance (McGrail 1978, i, 163-4, no. 19) of this identification adds greatly to the interest of the associated discoveries.

Excavation of the surrounding area revealed that the trough had apparently sunk in antiquity onto the edge of the lake-bed where it was firmly wedged between four upright posts; further posts were found flanking the trough which was interpreted as having been incorporated into the lower part of an artificial platform, the upper parts of which had been destroyed by fire. To the W of the boat there were found (Ward 1974, 21-5) the remains of what was apparently a 'small artificial box' which had suffered from burning. Set into this there was a 'box' (variously described as trapezoidal and rectangular) which had been carefully constructed from moss- and clay-caulked birch-poles. The 'box' was apparently watertight and had been filled with small pieces of burnt sandstone. It was initially interpreted as a possible fish-tank, but Bellhouse (1979) has cast doubt on both the interpretation and date of the object, seeing the log itself as a piece of re-used bog oak of considerably earlier date, and retaining its bark. The location is noted as not being a lake of navigable size, while lacustrine deposits were not revealed by geological sampling. The discovery is re-interpreted as that of a bothy and work-site of comparatively recent date, based on a perpetual supply of seepage-water for steeping or soaking, and a supply of poles. Skin-dressing, flax-retting and basket-making are all considered possible explanations. The discovery might also be explained as that of a logboat converted into a trough along the lines of that from Curraghtarsna (section IV.9.4, above).

*IV.9.7. SLEDGES AND SLIDES*

A typology of the sledge in Arctic and Nordic contexts has been developed by Berg (1935, *passim*.) who stresses its close relationship with the boat, seeing a common origin (possibly as early as the Mesolithic) in the primitive boat-sledge. This type developed, aided by the development of dog and reindeer haulage, into a variety of types including double-sledges and the single-runner monoxylous sledge that was used, typically for the carriage of manure, on clayey soil as well as on snow and ice. From this developed the Lapp boat-sledge with plank-built sides which keep the load above the mud. He cites and illustrates (pl. II and fig. 8) which it would appear difficult to distinguish from logboats in many, if not most, archaeological contexts, although the sledges appear to be generally smaller and considerations of stability may make their form distinct.

The construction and use of one or more distinct types of sledge or, more probably, hybrid sledge/boat/slide vehicle of monoxylous construction is highly probable at some period in Scotland, and the various suggestions that have been made regarding the use of the Eadarloch logboat (no. 36) as a timber-slide must be considered with that in mind. The use of such vehicles on land and water need not be mutually exclusive.

Fenton (1976, 200-6, *passim*.) notes the prevalence of sledges, sleds and travois-type vehicles for use in Scotland on dry and hard ground away from roads. The frame-built slipe was typical but sledges were used to carry such bulky loads as peat, dung and corn across moderately level ground; many of these may have been of monoxylous construction, particularly in the less-developed areas.

*IV.10. Conclusions*

The conscientious reader will by now have come to appreciate both the quantity and the limitations for logboats in Scotland, and, by implication, the potential of the country for further discoveries and research. Hopefully, the significance of the logboat in Scottish archaeological history will have been made apparent, as will the importance of the type within wetland archaeology in general. If this appreciation should further the progress of research, the present author will be well content.

*IV Synthesis and Analysis*

'Behold the hour, the boat arrive;
 Thou goest, the darling of my heart;
 Sever'd from thee, can I survive?
 But Fate has will'd and we must part.'

Robert Burns

# V Bibliography and Abbreviations

Abercromby, J 1905 'Report on Excavations at Fethaland and Trowie Knowe, Shetland; and of the Exploration of a Cairn on Dumglow, one of the Cleish Hills, Kinross-shire', *PSAS*, xxxix (1904-5), 171-84.

*AC Ayrshire Collections* (variously *Archaeological and Historical collections Relating to the Counties of Ayr and Wigton, Archaeological and Historical Collections Relating to Ayrshire and Galloway, Collections of the Ayrshire Archaeological and Natural History Society* and *Ayrshire Archaeological and Natural History Collections*).

Adam, J 1866 'Account of a Canoe of Oak found in the Castle Loch of Closeburn, Dumfriesshire', *PSAS*, vi (1864-6), 458.

Affleck, J 1912 'The Crannogs in Carlingwark Loch', *TDGNHAS*, 2nd series, xxiv (1911-12), 235-40.

Allen, D 1994 'Hot water and plenty of it', *Archaeology Ireland*, 8 (1994), 8-9.

Allen, J Romilly and Anderson, J 1903 *The Early Christian Monuments of Scotland*, Edinburgh.

Andersen, SH 1985 'Tybrind Vig: A Preliminary Report on a Submerged Ertebolle Settlement on the West Coast of Fyn', *Journal of Danish Archaeology*, 4 (1985), 52-69.

Andersen, SH 1986 'Mesolithic Dug-outs and Paddles from Tybrind Vig, Denmark', *Acta Archaeologica*, 57 (1986), 87-106.

Anderson, J 1883 *Scotland in Pagan Times: The Iron Age*, Edinburgh.

Anderson, J 1885 'Notice of a Bronze Cauldron found with several Kegs of Butter in a Moss near Kyleakin, Skye; with notes of other Caldrons of Bronze found in Scotland', *PSAS*, xix (1884-5), 309-15.

Anderson, J 1886 *Scotland in Pagan Times: The Bronze and Stone Ages*, Edinburgh.

Anderson, ML 1967 *A history of Scottish forestry*, (Taylor, CJ ed.), London.

Andrian, B 1995 'Scottish crannog centre - opening soon!', *Newswarp: the newsletter of the Wetland Archaeological research Project,* no. 18 (November 1995), 3-5.

ANGMAG F Angus District Museum, Library and Museum, High Street, Forfar.

Arnold, B 1978 'Gallo-Roman boat finds in Switzerland' in Taylor, J du P and Cleere, H (eds.) *Roman shipping and trade: Britain and the Rhine provinces*, Council for British Archaeology Research Report series no. 24, 31-5, London.

Arnold, B 1982 'The architectural woodwork of the Late Bronze Age village Auvernier-Nord' in McGrail, S (ed.) *Woodworking Techniques before A.D. 1500: Papers presented to a Symposium at Greenwich in September, 1980, together with edited discussion*, British Archaeological Reports international series no. 129 and National Maritime Museum archaeological series no. 7, 111-28, Oxford.

Ashbee, P 1957 'The Great Barrow at Bishop's Waltham, Hampshire', *PPS*, new series, xxiii (1957), 137-66.

AUAM Aberdeen University Anthropological Museum, Marischal College, Aberdeen AB9 1AS.

Baillie, MGL 1977 'An oak chronology for south central Scotland', *Tree-ring Bulletin*, 37 (1977), 33-44.

Balfour, JA (ed.) *The Book of Arran: Archaeology*, Glasgow.

Barber, J 1982 'A wooden bowl from Talisker, Skye', *PSAS*, 112 (1982), 578-9.

Barber, J 1984 'Medieval wooden bowls' in Breeze, DJ (ed.) *Studies in Scottish antiquity, presented to Stewart Cruden*, Edinburgh, 125-47.

Barfield, L and Hodder, M 1987 'Burnt mounds as saunas: an exercise in archaeological interpretation', *Antiquity*, 61 (1987), 370-9.

Barry, TB 1988 *The Archaeology of Medieval Ireland*, paperback ed., London.

Begg, RB 1888 'Notice of a Crannog discovered in Lochleven, Kinross-shire, on 7th September 1887', *PSAS*, xxii (1887-8), 118-24.

Bellhouse, RL 1979 'The Branthwaite boat', *Council for British Archaeology, Regional Group 3 Newsletter*, (1979), 13-16.

Berg, G 1935 *Sledges and wheeled vehicles: Ethnological studies from the view-point of Sweden*, Stockholm, Sweden.

Bersu, G and Wilson, DM 1966 *Three Viking graves in the Isle of Man*, Society for Medieval Archaeology monograph series, no. 1, London.

Black, GF 1904 'Descriptive Catalogue of Antiquities found in Ayrshire and Wigtownshire, and now in the National Museum, Edinburgh', *AC*, vii (1894), 1-47.

Blundell, FO 1911 'Notes on the Church and some Sculptured Monuments in the Churchyard at Saint Maelrubha in Arisaig, and on an Artificial Island there; also on some Sculptured Monuments in the Churchyard of Kilchoan, Knoydart', *PSAS*, xlv (1910-11), 353-66.

Boe, G de 1978 'Roman boats from a small river harbour at Pommeroeul, Belgium' in Taylor, J du P and Cleere, H (eds.) *Roman shipping and trade: Britain and the Rhine provinces*, Council for British Archaeology Research Report series no. 24, London, 22-30.

Brindley, AL, Lanting, JL and Mook, WG 1990 'Radiocarbon Dates from Irish Fulachta Fiadh and Other Burnt Mounds', *Journal of Irish Archaeology*, v (1989-90), 25-33.

Brown, J 1891 *The History of Sanquhar*, 2nd ed., Dumfries.

Bruce, J 1893 *History of the Parish of West or Old Kilpatrick, and of the church and certain lands in the parish of east or New Kilpatrick*, Glasgow.

Bruce, J 1900 'Notes on the Discovery and exploration of a Pile Structure on the North Bank of the River Clyde, east from Dumbarton Rock', *PSAS*, xxxiv (1899-1900), 437-62.

[Buchanan. J] 1848 'Canoes, Ancient, found at Glasgow', *Edinburgh Topographical, Traditional and Antiquarian Magazine*, (1848), 168-74.

Buchanan, J 1854a 'Discovery of Ancient Canoes on the Clyde', *PSAS*, i (1851-4), 44-5.

Buchanan, J 1854b 'Notice of the Discovery of an Ancient Boat, of Singular Construction, on the Banks of the Clyde', *PSAS*, i (1851-4), 211-13.

Buchanan, J 1883 'Address to the Society at its annual meeting on 2nd February, 1869', *Transactions of the Glasgow Archaeological Society*, ii (1883), 66-77.

[Buchanan, J] 1884 'Ancient canoes found at Glasgow' in 'Senex', *Glasgow Past and present*, ii, Glasgow, 342-67.

Buckley, V 1985 'Curraghtarsna', *Current Archaeology*, no. 98 (1985), 70-1.

Bulleid, A 1949 *The Lake-Villages of Somerset*, 4th ed., Yeovil.

Burgess, C 1980 *The Age of Stonehenge*, London.

Burnett, JH 1854 'Bronze Vessels Discovered in the Loch of Leys', *PSAS*, i (1851-4), 26-7.

Burns, J 1972 'A Wooden Boat and Associated Structures, Branthwaite, Workington, Cumbria (NY 056 247)', *Council for British Archaeology, Regional Group 3 Newsletter*, (1972), 6-7.

Burns-Begg, R 1887 *History of Lochleven Castle*, 2nd ed., Kinross.

Burns-Begg, R 1901 *The secrets of my prison-house, Being full Details of Queen Mary's Experiences in Lochleven Castle*, revised ed., Kinross.

Calder, CST 1950 'Report on the Excavation of a Neolithic Temple at Stanydale in the Parish of Sandsting, Shetland', *PSAS*, lxxxiv (1949-50), 182-205.

Caldwell, DH 1981 'Some Notes on Scottish Axes and Long Shafted Weapons' in Caldwell, DH (ed.) *Scottish Weapons and Fortifications, 1100-1800*, Edinburgh, 253-314.

Callander, JG 1929 'Land Movements in Scotland in Prehistoric and Recent Times', *PSAS*, lxiii (1928-9), 314-22.

Campbell, F 1877 'Notice of an Artificial Island, and an Ancient Canoe, found in Draining a Loch near Tobermory, Mull', *PSAS*, viii (1868-70), 465.

Campbell, M and Sandeman, M 1962 'Mid Argyll: an Archaeological Survey', *PSAS*, xcv (1961-2), 1-125.

Cathcart, FM 1857 'Account of the Discovery of a number of Ancient Canoes of solid oak, in Loch Doon, a fresh-water Lake in the county of Ayr', *Archaeologia Scotica*, iv (1857), 299-301.

Challis, AJ and Harding, DW 1975 *Later Prehistory from the Trent to the Tyne*, Oxford.

Chalmers, P 1844-59 *Historical and Statistical Account of Dunfermline*, Edinburgh.

Chapman, H 1982 'Roman vehicle construction in the northwest provinces' in McGrail, S (ed.) *Woodworking Techniques before A.D. 1500: Papers presented to a Symposium at Greenwich in September, 1980, together with edited discussion*, British Archaeological Reports international series no. 129 and National Maritime Museum archaeological series no. 7, Oxford, 187-93.

Chapman, RW 1924 (ed.) *Johnson's Journey to the Western Islands of Scotland and Boswell's Journal of a Tour to the Hebrides with Samuel Johnson, LL.D.*, Oxford.

Cheape, H 1993 'Woodlands on the Clanranald estate: a case study' in Smout, TC (ed.) *Scotland since Prehistory: Natural Change and Human Impact*, Aberdeen.

Childe, VG 1935 *The prehistory of Scotland*, London.

Childe, VG 1940 *Prehistoric communities of the British Isles*, London.

Childe, VG 1946 *Scotland before the Scots, being the Rhind lectures for 1944*, London.

Chiltern Seeds 1986 *Grow Something New From Seed*, Catalogue of Chiltern Seeds, Bortree Stile, Ulverston, Cumbria.

Christensen, C 1990 'Stone Age Dug-Out Boats in Denmark: Occurrence, Age, Form and Reconstruction' in Robinson, DE (ed.) *Experimentation and Reconstruction in Environmental Archaeology*, (Symposia of the Association for Environmental Archaeology No. 9, Roskilde, Denmark, 1988), Oxford, 119-41.

Christensen, AE 1982 'Viking Age boatbuilding tools' in McGrail, S (ed.) *Woodworking Techniques before A.D. 1500: Papers presented to a Symposium at Greenwich in September, 1980, together with edited discussion*, British Archaeological Reports international series no. 129 and National Maritime Museum archaeological series no. 7, Oxford, 327-37.

Christison, R 1881 'On an Ancient Wooden Image, found in November last at Ballachulish Peat-Moss', *PSAS*, xv (1880-1), 158-78.

Clark, JGD and Godwin, H 1940 'A Late Bronze Age Find near Stuntney, Isle of Ely', *Antiquaries' Journal*, xx (1940), 52-71.

Clarke, DV 1981 'Scottish Archaeology in the Second Half of the Nineteenth Century' in Bell, AS (ed.) *The Scottish Antiquarian Tradition: Essays to mark the bicentenary of the Society of Antiquaries of Scotland and its Museum, 1780-1980*, Edinburgh, 114-41.

Clarke, DV and Sharples, N 1985 'Settlements and Subsistence in the Third Millennium BC' in Renfrew, C (ed.) *The Prehistory of Orkney*, Edinburgh, 54-82.

Clarke, DV, Breeze, DJ and Mackay, G 1980 *The Romans in Scotland: An introduction to the collections of the National Museum of Antiquities of Scotland*, Edinburgh.

Close-Brooks, J 1975 'An iron-age date for the Loch Lotus canoe', *PSAS*, 106 (1974-5), 199.

Close-Brooks, J 1984 'Some objects from peat bogs', *PSAS*, 114 (1984), 578-81.

Cochrane Patrick, RW 1872 'Notices of some Antiquities recently observed in North Ayrshire', *PSAS*, ix (1870-2), 385-7.

Coles, FR 1984 'The Motes, Forts and Doons in the East and West Divisions of the Stewartry of Kirkcudbright', *PSAS*, xxvii (1892-3), 92-182.

Coles, B and J 1986 *Sweet track to Glastonbury: The Somerset Levels in Prehistory*, London.

Coles, J [M] 1973 *Archaeology by experiment*, London.

Coles, JM 1982 'Ancient woodworking techniques: the implications for archaeology' in McGrail, S (ed.) *Woodworking Techniques before A.D. 1500: Papers presented to a Symposium at Greenwich in September, 1980, together with edited discussion*, British Archaeological Reports international series no. 129 and National Maritime Museum archaeological series no. 7, Oxford, 1-6.

Coles, J [M] 1984 *The Archaeology of Wetlands*, Edinburgh.

Coles, JM 1988 'The peat hag' (publication of the 12th Beatrice de Cardi lecture), *Council for British Archaeology Report No. 38 for the year ended 30 June 1988*, London, 68-73.

Coles, JM, Heal, SVE and Orme, BJ 1978 'The use and character of wood in prehistoric Britain and Ireland', *PPS*, 44 (1978), 1-45.

Coles, JM and Harding, AF 1979 *The Bronze Age in Europe: An introduction to the prehistory of Europe c.2000-750 BC*, London.

Coles, J [M] and Minnitt, S 1995 *'Industrious and fairly civilized': The Glastonbury lake village*, [Taunton].

Coppock, JT 1976 *An Agricultural Atlas of Scotland*, Edinburgh.

Corrie, JM 1928 'Kirkcudbright in the Stone, Bronze and Iron Ages', *TDGNHAS*, 3rd series, xiv (1926-8), 272-99.

Coutts, H 1970 *Ancient Monuments of Tayside*, Dundee.

Coutts, H 1971 *Tayside Before History: A guide-catalogue of the collection of antiquities in the Dundee Museum*, Dundee.

Crone, [B] A 1991 'Buiston crannog', *Current Archaeology*, no. 127 (1991), 295-7.

Crone, BA 1993a 'Crannogs and chronologies', *PSAS*, 123 (1993), 245-54.

Crone, BA 1993b 'A wooden bowl from Loch a'Ghlinne Bhig, Bracadale, Skye', *PSAS*, 123 (1993), 269-75.

Crone, [B] A and Barber, J 1981 'Analytical techniques for the investigation of non-artifactual wood from prehistoric and medieval sites', *PSAS*, 111 (1981), 510-15.

Crumlin-Pedersen, O 1972 'Skin or wood? A study of the origin of the Scandinavian plank-boat' in Hasslof, O *et al. Ships and shipyards, sailors and fishermen*, Copenhagen, Denmark.

Cullingford, RA, Caseldine, CJ and Gotts, PE 1980 'Early Flandrian land and sea-level changes in Lower Strathearn', *Nature*, 284 (13 March 1980), 159-61.

Curle, CL 1982 *Pictish and Norse finds from the Brough of Birsay 1934-74*, Society of Antiquaries of Scotland monograph series no. 1, Edinburgh.

Curle, J 1911 *A Roman frontier post and its people: the fort of Newstead in the parish of Melrose*, Glasgow.

Cursiter, JW 1887 'Notice of a Canoe recently found in the Island of Stronsay, Orkney', *PSAS*, xxi (1886-7), 279-81.

Curwen, EC 1940 'A Viking Ship-burial at Stranraer', *Antiquity*, xiv (1940), 434.

Curwen, EC 1954 *The archaeology of Sussex*, 2nd ed., London.

Dalgarno, J 1876 *Notes on the Parishes of Slains and Forvie, in the Olden Days*, Aberdeen.

Dalrymple, CE 1872 'Notes of the Examination of a Crannog in the Black Loch, anciently called "Loch Inch-Cryndil," Wigtownshire', *PSAS*, ix (1870-2), 388-92.

Daniel, G 1981 *A Short History of Archaeology*, London.

Darbishire, RD 1874 'Notes on Discoveries in Ehenside Tarn, Cumberland', *Archaeologia*, xliv (1874), 273-92.

Darrah, R 1982 'Working unseasoned oak' in McGrail, S (ed.) *Woodworking Techniques before A.D. 1500: Papers presented to a Symposium at Greenwich in September, 1980, together with edited discussion*, British Archaeological Reports international series no. 129 and National Maritime Museum archaeological series no. 7, Oxford, 219-29.

Darvill, T 1987 *Ancient monuments in the countryside: an archaeological management review*, English Heritage Archaeological Report no. 5, London.

Davidson, DA and Jones, RL 1985 'The Environment of Orkney' in Renfrew, C (ed.) *The Prehistory of Orkney*, Edinburgh, 10-25.

*DES* (Date) *Discovery and Excavation in Scotland*, formerly *Discovery and Excavation, Scotland*, annual publication of the Council for Scottish Archaeology (formerly Scottish Group and previously Scottish Regional Group), Council for British Archaeology.

Dixon, JS 1875 'Notes on the discovery of an Ancient Canoe on the farm of Littlehill, Cadder Moor, near Kirkintilloch, the property of Sir William Stirling Maxwell, Bart.', *Proceedings of the Natural History Society of Glasgow*, ii (1869-75), 65-6.

Dixon, TN 1981 'Preliminary excavation of Oakbank crannog, Loch Tay: Interim report', *International Journal of Nautical Archaeology and Underwater Excavation*, 10 (1981), 15-21.

Dixon, [T] N 1982a 'Excavation of Oakbank crannog, Loch Tay: Interim report', *International Journal of Nautical Archaeology and Underwater Excavation*, 11 (1982), 125-32.

Dixon, TN 1982b 'A survey of crannogs in Loch Tay', *PSAS*, 112 (1982), 17-38.

Dixon, [T] N 1984 'Oakbank crannog', *Current Archaeology*, no. 90 (1984), 217-20.

DMAG Dundee Museums and Art Galleries, McManus Galleries, Albert Square, Dundee DD1 1DA.

Dodwell, C 1989 *Travels with Pegasus: A microlight journey across West Africa*, London.

DUMFM Dumfries Museum (formerly Dumfries Burgh Museum), The Observatory, Church Street, Dumfries DG2 7SW.

Duncan, JD 1883 'Note regarding the ancient canoe recently discovered in the bed of the Clyde above the Albert Bridge', *Transactions of the Glasgow Archaeological Society*, ii (1883), 121-30.

Duns, J 1883 'Notes on North Mull (Second Communication)', *PSAS*, xvii (1882-3), 337-50.

Earwood, C 1988 'Wooden containers and other wooden artifacts from the Glastonbury lake village', *Somerset Levels Papers*, 14 (1988), 82-93.

Earwood, C 1990a 'The wooden artefacts from Loch Glashan crannog, Mid Argyll', *PSAS*, 120 (1990), 79-94, fiche 1: E1-8.

Earwood, C 1990b 'Radiocarbon Dating of Late Prehistoric Wooden Vessels', *Journal of Irish Archaeology*, v (1989-90), 27-44.

Earwood, C 1991 'Two Early Historic bog butter containers', *PSAS*, 121 (1991), 231-40.

Earwood, C 1992 'A Radiocarbon Date for Early Bronze Age Wooden Polypod Bowls', *Journal of Irish Archaeology*, vi (1991-2), 27-8.

Earwood, C 1993a *Domestic Wooden Artefacts in Britain and Ireland from Neolithic to Viking times*, Exeter.

Earwood, C 1993b 'The dating of wooden troughs and dishes', *PSAS*, 123 (1993), 355-62.

Elgee, HW and F 1949 'An Early Bronze Age Burial in a Boat-shaped Wooden Coffin from North-east Yorkshire', *PPS*, new series, xv (1949), 87-106.

Ellmers, D 1978 'Shipping on the Rhine during the Roman period: the pictorial evidence' in Taylor, J du P and Cleere, H (eds.) *Roman shipping and trade: Britain and the Rhine provinces*, Council for British Archaeology Research Report series no. 24, London, 1-14.

Ellmers, D 1981 'Post-Roman waterfront installations on the Rhine' in Milne, G and Hobley, B (eds.) *Waterfront archaeology in Britain and northern Europe*, Council for British Archaeology Research Report series no. 41, London, 88-95.

Evans, J 1897 *The ancient stone implements, weapons and ornaments, of Great Britain*, 2nd ed., London.

Fairhurst, H 1969 'A mediaeval island-settlement in Loch Glashan', *Glasgow Archaeological Journal*, 1 (1969), 47-67.

Feachem, RW 1959 'A Dug-out Canoe from Cambuskenneth Abbey', *PSAS*, xcii (1958-9), 116-17.

Feachem, RW 1977 *Guide to Prehistoric Scotland*, 2nd ed., London.

Fenton, A 1963 'Early and Traditional Cultivating Implements in Scotland', *PSAS*, xcvi (1962-3), 264-317.

Fenton, A 1972 'The Currach in Scotland, with Notes on the Floating of timber', *Scottish Studies*, 16 (1972), 61-81.

Fenton, A 1976 *Scottish Country Life*, Edinburgh.

Fenton, A 1978 *The Northern Isles: Orkney and Shetland*, Edinburgh.

Fenton, A 1984a 'Wheelless Transport in Northern Scotland' in Fenton, A and Stell, G (eds.) *Loads and Roads in Scotland and Beyond: Land Transport over 6000 Years*, Edinburgh, 105-23.

Fenton, A 1984b 'The Distribution of Carts and Wagons' in Fenton, A and Stell, G (eds.) *Loads and Roads in Scotland and Beyond: Land Transport over 6000 Years*, Edinburgh, 124-40.

Fenton, A and Walker, B 1981 *The Rural Architecture of Scotland*, Edinburgh.

Fleming, JS 1915 'Notes on the Remains of a Crannog in Loch Vennachar', *PSAS*, xlix (1914-15), 341-2.

Forsyth, R 1805-8 *The Beauties of Scotland: Containing a Clear and Full Account of the Agriculture, Commerce, Mines and Manufactures; of the Population, Cities, Towns, Villages, &c. of Each County*, Edinburgh.

Fox, C 1926 'A Dug-out Canoe from South Wales: with Notes on the Chronology, Typology, and Distribution of Monoxylous Craft in England and Wales', *Antiquaries' Journal*, vi (1926), 121-51.

Fraser, HA 1917 'Investigation of the Artificial Island in Loch Kinellan, Strathpeffer', *PSAS*, li (1916-17), 48-98.

Fraser, H [A] 1918 'Artificial Islands in the Dingwall District', *Transactions of the Inverness Scientific Society and Field Club*, viii (1912-18), 231-62.

Fry, MF 1988 *Paddle your own! The logboat in the North of Ireland*, publication of the Historic Monuments and Buildings Branch, Department of the Environment (Northern Ireland), Belfast.

GAGM Glasgow Art Gallery and Museum, Kelvingrove, Glasgow G3 8AG.

Garner, A, Prag, J and Housley, R 1994 'The Alderley Edge Shovel', *Current Archaeology*, no. 137 (1994), 172-5.

Geddes, J 1982 'The construction of medieval doors' in McGrail, S (ed.) *Woodworking Techniques before A.D. 1500: Papers presented to a Symposium at Greenwich in September, 1980, together with edited discussion*, British Archaeological Reports international series no. 129 and National Maritime Museum archaeological series no. 7, Oxford, 313-25.

Geikie, J 1877 *The Great Ice Age And Its Relation to the Antiquity of Man*, 2nd ed., London.

Geikie, J 1880 'Discovery of an ancient Canoe in the old alluvium of the Tay at Perth', *The Scottish Naturalist: A Magazine of Natural History*, v (1879-80), 1-7.

Geikie, J 1881 *Prehistoric Europe: a Geological Sketch*, London.

Geikie, J 1894 *The Great Ice Age and its Relation to the Antiquity of man*, 3rd ed., London.

Gifford, E 1993 'Expanding oak logboats: is it possible?' in Coles, J [M], Fenwick, V and Hutchinson, G (eds.) *A Spirit of Enquiry: Essays for Ted Wright*, Exeter, 52-3.

Gillespie, JE 1876 'Notice of a Canoe found in Loch Lotus, Parish of New Abbey, Kirkcudbrightshire', *PSAS*, xi (1874-6), 21-3.

Gillespie, R 1876 *Glasgow and the Clyde*, Glasgow.

Glob, PV 1974 *The Mound People: Danish Bronze-Age Man Preserved*, London.

Godsman, J [1952] *King-Edward, Aberdeenshire, the story of a parish*, Banff.

Gomme, GL (ed.) 1886 *The Gentleman's Magazine Library: being a classified collection of the chief contents of the Gentleman's Magazine from 1731 to 1868, Archaeology*, London.

Good, GL and Tabraham, CJ 1981 'Excavations at Threave Castle, Galloway, 1974-8', *Medieval Archaeology*, 25 (1981), 90-140.

Goodburn, D and Redknap, M 1988 'Replicas and wrecks from the Thames area', *The London Archaeologist*, 6.1 (Winter 1988), 7-10, 19-22.

Goodman, WL 1964 *The History of Woodworking Tools*, London.

Graham, JM 1982 'Quantitative Methods and Boat Archaeology' in McGrail, S (ed.) *Woodworking Techniques before A.D. 1500: Papers presented to a Symposium at Greenwich in September, 1980, together with edited discussion*, British Archaeological Reports international series no. 129 and National Maritime Museum archaeological series no. 7, Oxford, 137-55.

Graham-Campbell, J 1980 *Viking Artefacts: A Select Catalogue*, London.

Green, HS 1987 'The Disgwylfa Fawr Round Barrow, Ceredigion, Dyfed', *Archaeologia Cambrensis*, cxxxvi (1987), 43-50.

Green, S 1985 'The Caergwrle Bowl - not oak but shale', *Antiquity*, lix (1985), 116-17.

Greenhill, B 1976 *Archaeology of the boat: A new introductory study*, London.

Grigor, J 1864a 'Notice of the Remains of Two Ancient Lake Dwellings in the Loch of the Clans, on the Estate of James Rose, of Kilravock, Esq., Nairnshire, with a Plan', *PSAS*, v (1862-4), 116-19.

Grigor, J 1864b 'Further Explorations of the Ancient Lake Dwellings in the Loch of the Clans, on the Estate of Kilravock, Nairnshire', *PSAS*, v (1862-4), 332-5.

Guido, [C] M 1974 'A Scottish crannog re-dated', *Antiquity*, xlvii (1974), 54-5.

Haddon, AC 1937 'The Canoes of Melanesia, Queensland and New Guinea', vol. ii of Haddon, AC and Hornell, J *Canoes of Oceania*, Bernice P Bishop Museum special publication no. 28, Honolulu, Hawaii, USA.

Hale, JR 1980 'Plank-built in the Bronze Age', *Antiquity*, liv (1980), 118-26.

Hall, RA 1982 '10th century woodworking in Coppergate, York' in McGrail, S (ed.) *Woodworking Techniques before A.D. 1500: Papers presented to a Symposium at Greenwich in September, 1980, together with edited discussion*, British Archaeological Reports international series no. 129 and National Maritime Museum archaeological series no. 7, Oxford, 231-44.

Hanson, WS 1978 'The Organisation of Roman Military Timber-Supply', *Britannia*, ix (1978), 293-305.

Hanson, WS 1982 'Roman military timber buildings: construction and reconstruction' in McGrail, S (ed.) *Woodworking Techniques before A.D. 1500: Papers presented to a Symposium at Greenwich in September, 1980, together with edited discussion*, British Archaeological Reports international series no. 129 and National Maritime Museum archaeological series no. 7, Oxford, 169-86.

Hanson, WS and Maxwell, GS 1983 *Rome's North West Frontier: the Antonine Wall*, Edinburgh.

Hardy, BL, McArdle, DT, Miles, DL and Morrison, IA [1972] *A report on the survey of Loch Awe for evidence of lake dwellings and a mollusc survey for the British Conchological Society*, Naval Air Command Sub Aqua Club typescript report.

Harper, M M' L 1876 *Rambles in Galloway: Topographical, Historical, Traditional and Biographical*, Edinburgh.

Harrison, RJ 1980 *The Beaker Folk: Copper Age archaeology in Western Europe*, London.

Heal, SVE 1982 'The Wood Age? The significance of wood usage in pre-Iron Age north western Europe' in McGrail, S (ed.) *Woodworking Techniques before A.D. 1500: Papers presented to a Symposium at Greenwich in September, 1980, together with edited discussion*, British Archaeological Reports international series no. 129 and National Maritime Museum archaeological series no. 7, Oxford, 169-86.

Heal, [S] V [E] 1986 'Comment on the form of the central grave coffin', specialist report in Lawson, AJ 'The Excavation of a Ring-ditch at Bowthorpe, Norwich, 1979' in Lawson, AJ (ed.) *Barrow Excavations in Norfolk, 1950-82*, East Anglian Archaeology, report no. 29, Dereham, 45-8.

Heal, SVE and Hutchinson, G 1986 'Three recently found logboats', *International Journal of Nautical Archaeology and Underwater Exploration*, 15 (1986), 205-13.

Hedges, J 1975 'Excavation of two Orcadian burnt mounds at Liddle and Beaquoy', *PSAS*, 106 (1974-5), 39-98.

Hedges, REM, Housley, RA, Bronk, CR and van Klinken, GJ 1991 'Radiocarbon dates from the Oxford AMS system: Archaeometry datelist 12', *Archaeometry*, 33 (1991), 121-34.

Hedges, REM, Housley, RA, Bronk, CR and van Klinken, GJ 1992 'Radiocarbon dates from the Oxford AMS system: Archaeometry datelist 14', *Archaeometry*, 34 (1992), 141-59.

Hedges, REM, Housley, RA, Bronk-Ramsey, C and van Klinken, GJ 1993 'Radiocarbon dates from the Oxford AMS system: Archaeometry datelist 16', *Archaeometry*, 35 (1993), 147-67.

Hencken, H 1951 'Lagore crannog: an Irish royal residence of the 7th to 10th centuries A.D.', *Proceedings of the Royal Irish Academy*, 53, section C (1950-1), 1-247.

Henderson, E 1990 *A Parish Alphabet, Being for the most part an account of the topographical features of the parish of Ballingry*, Troon.

Hewett, CA 1980 *English historic carpentry*, London.

Hewett, CA 1982 'Toolmarks on surviving works from the Saxon, Norman and later medieval period' in McGrail, S (ed.) *Woodworking Techniques before A.D. 1500: Papers presented to a Symposium at Greenwich in September, 1980, together with edited discussion*, British Archaeological Reports international series no. 129 and National Maritime Museum archaeological series no. 7, Oxford, 339-48.

Hewison, JK 1939 *The Romance of Dumfries and Galloway in Early Caledonia*, Dumfries.

HM Hunterian Museum, University of Glasgow, Glasgow G12 8QQ.

Hobley, B 1981 'The London Waterfront - the exception or the rule?' in Milne, G and Hobley, B (eds.) *Waterfront archaeology in Britain and northern Europe*, Council for British Archaeology Research Report series no. 41, London, 1-9.

Hodges, H 1964 *Artifacts: an introduction to early materials and technology*, London.

Hogg, A 1890 'The Antiquities of Davan and Kinnord', *The Scottish Naturalist*, new series, iv (1889-90), 157-71.

Hornell, J 1950 *Fishing in Many Waters*, Cambridge.

Hornell, J 1970 *Water Transport: Origin and Evolution*, reprint of 1946 ed., Newton Abbot.

HS Historic Scotland (formerly Historic Buildings and Monuments and other prior designations), Longmore House, Salisbury, Edinburgh EH9 1SH.

Humphries, CJ Press, JR and Sutton, DA 1981 *The Hamlyn Guide to Trees of Britain and Europe*, London.

Hurley, M 1982 'Wooden artifacts from the excavation of the medieval city of Cork' in McGrail, S (ed.) *Woodworking Techniques before A.D. 1500: Papers presented to a Symposium at Greenwich in September, 1980, together with edited discussion*, British Archaeological Reports international series no. 129 and National Maritime Museum archaeological series no. 7, Oxford, 307-11.

Hutcheson, A 1897 'Notices (1) of an Ancient Canoe found in the river Tay, near Errol; (2) a Grinding-stone found on the Sidlaw Hills; (3) a Beggar's Badge of Sixteenth Century found in Dundee; and (4) a Spear-head of Flint found in the carcass of a whale. Being recent additions to the Dundee Museum', *PSAS*, xxxi (1896-7), 265-81.

Hutchinson, G 1986 'The Southwold side rudders', *Antiquity*, lx (1986), 219-21.

INVMG Inverness Museum and Art Gallery, Castle Wynd, Inverness IV2 3ED.

Irving, J 1920 *Dumbartonshire: County and Burgh From the earliest times to the close of the Eighteenth Century, forming part II of a revised History of Dumbartonshire*, Dumbarton.

Jamieson, R (ed.) 1876 *Burt's Letters from the North of Scotland*, also known as *Letters from a Gentleman in the North of Scotland to his Friend in London*, 5th ed., London.

Jardine, W 1865 'Address of the President', *TDGNHAS*, 1st series, ii (1864-5), 1-24.

Jardine, WG and Masters, LS 1977 'A Dugout Canoe from Catherinefield Farm', *TDGNHAS*, 3rd series, lii (1976-7), 56-65.

Joass, JM 1881 'Note on the "Curach" and "Ammir" in Ross-shire', *PSAS*, xv (1880-1), 179-80.

Jobey, G 1984 'The Cartington coffin: a radiocarbon date', *Archaeologia Aeliana*, 5th series, xii (1984), 235-7.

Johnstone, P 1980 *The Sea-craft of Prehistory*, London.

Jolly, W 1876 'St Columba's Loch, in Skye, and its Ancient Canoes', *PSAS*, xi (1874-6), 551-61.

Joncheray, D 1986 'Les embarcations monoxyles dans la region pays de la Loire', *Études Préhistoriques et Historiques des Pays de la Loire*, 9 (1986). Publication of the Association d'Études Préhistoriques et Historiques des Pays de la Loire, Nantes, France.

Kemp, P (ed.) 1979 *The Oxford Companion to Ships and the Sea*, paperback ed., Oxford.

Kennedy, A 1831 'Notice respecting an Ancient Ship discovered in a Garden at Stranrawer, in Galloway', *Archaeologia Scotica*, iii (1831), 51-2.

Klindt-Jensen, O 1981 'Archaeology and Ethnography in Denmark: early studies' in Daniel, G (ed.) *Towards a History of Archaeology*, London, 14-19.

Lageard, JGA and Chambers, FM 1993 'The palaeological significance of a new subfossil-oak (*Quercus* sp.) chronology from Morris' Bridge, Shropshire, UK', *Dendrochronology*, 11 (1993), 25-33.

Laing, A 1876 *Lindores Abbey and its Burgh of Newburgh: their history and annals*, Edinburgh.

Lang, A 1905 *The Clyde Mystery: A Study in Forgeries and folklore*, Glasgow.

Lerche, G 1977 'Double paddle-spades in prehistoric contexts in Denmark', *Tools and Tillage*, iii, pt. 2 (1977), 111-24.

Love, R 1876 'Notices of the several Openings of a Cairn on Cuffhill; of various Antiquities in the Barony of Beith; and of a Crannog in the Loch of Kilbirnie, Ayrshire', *PSAS*, xi (1874-6), 272-97.

Lucas, AT 1963 'The Dugout Canoe in Ireland: The Literary Evidence', *Varbergs Museum Arsbok*, 14 (1963), 57-68.

Lyell, C 1829 'On a recent Formation of Freshwater Limestone in Forfarshire, and on some recent Deposits of freshwater Marl; with a Comparison of recent with ancient Freshwater Formations; and an Appendix on the Grygonite or Seed-vessel of the Chara', *Transactions of the Geological Society of London*, 2nd series, ii (1829), 73-96.

Lynn, CJ 1986 'Lagore, County Meath and Ballinderry No. 1, County Westmeath Crannogs: Some possible structural reinterpretations', *Journal of Irish Archaeology*, iii (1985-6), 69-73.

Macadam, WI 1882 'On the Results of a Chemical Investigation into the Composition of the "Bog Butters", and of "Adipocere" and the "Mineral Resins", with Notice of a Cask of Bog Butter found in Glen Gell, Morvern, Argyllshire, and now in the Museum', *PSAS*, xvi (1881-2), 204-23.

Macadam, WI 1889 'Notes of the Analysis of Additional Samples of Bog Butter found in different parts of Scotland', *PSAS*, xxiii (1888-9), 433-4.

Macalister, RAS 1928 *The archaeology of Ireland*, London.

McArdle, CM and TD 1973 'Loch Awe Crannog Survey', *The Kist: The Magazine of the Natural History and Antiquarian Society of Mid-Argyll*, 5 (1973), 2-12.

McCaig, E 1954 ' "Canoe" from Piltanton Burn', *TDGNHAS*, 3rd series, xxxii (1953-4), 178-9.

MacCartney, WN 1869 'On Ancient Canoes recently found at Renfrew', *Proceedings of the Natural History Society of Glasgow*, i (1858-69), 168.

MacDonald, A 1941 *The Place-names of West Lothian*, Edinburgh and London.

Macdonald, J 1882 'Illustrated Notices of the Ancient Stone Implements of Ayrshire (First Series)', *AC*, iii (1882), 66-81.

M'Dowall, W 1909 'Notice of a Canoe found at Kirkmahoe', *TDGNHAS*, 3rd series, vii (1919-20), 9-10.

Macfarlane, W 1906-8 *Geographical Collections relating to Scotland*, Mitchell, A and Clark, JT (eds.) Edinburgh.

MacGeorge, A 1880 *Old Glasgow, the Place and the People*, Glasgow.

McGrail, S 1975 'The Brigg Raft Re-excavated', *Lincolnshire History and Archaeology*, 10 (1975), 5-13.

McGrail, S 1977 'Searching for pattern among the Logboats of England and Wales' in McGrail, S (ed.) *Sources and techniques in Boat Archaeology: Papers based on those presented to a Symposium at Greenwich in September 1976, together with edited discussion*, British Archaeological Reports supplementary series no. 29 and National Maritime Museum, Greenwich, archaeological series no. 2, Oxford, 115-35.

McGrail, S 1978 *Logboats of England and Wales with comparative material from European and other countries*, British Archaeological Reports British series no. 51 and National Maritime Museum, Greenwich, archaeological series no. 2, Oxford.

McGrail, S 1979 'Prehistoric Boats, Timber and Woodworking Technology', *PPS*, 45 (1979), 159-63.

McGrail, S 1981a 'A medieval logboat from the R. Calder at Stanley Ferry, Wakefield, Yorkshire', *Medieval Archaeology*, 25 (1981), 160-4.

McGrail, S 1981b 'Medieval boats, ships and landing places' in Milne, G and Hobley, B (eds.) *Waterfront archaeology in Britain and northern Europe*, Council for British Archaeology Research Report series no. 41, London, 17-23.

McGrail, S 1981c *The Ship: Rafts, Boats and Ships From Prehistoric Times to the Medieval Era*, London.

McGrail, S 1985a 'Towards a classification of water transport', *World Archaeology*, 16 pt. 3 (1985), 289-303.

McGrail, S 1985b 'The Hasholme Logboat', *Antiquity*, lix (1985), 117-20.

McGrail, S 1987a *Ancient Boats in N. W. Europe: The archaeology of water transport to AD 1500*, London.

McGrail, S 1987b 'Early boatbuilding techniques in Britain - dating technological change', *International Journal of Nautical Archaeology and Underwater Exploration*, 16 (1987), 343-54.

McGrail, S 1988 'Assessing the performance of an ancient boat - the Hasholme logboat', *Oxford Journal of Archaeology*, 7.1 (March 1988), 35-46.

McGrail, S and Denford, G 1982 'Boatbuilding techniques, technological change and attribute analysis' in McGrail, S (ed.) *Woodworking Techniques before A.D. 1500: Papers presented to a Symposium at Greenwich in September, 1980, together with edited discussion*, British Archaeological Reports international series no. 129 and National Maritime Museum archaeological series no. 7, Oxford, 25-72.

McGrail, S and Millett, M 1985 'The Hasholme logboat', *Antiquity*, lix (1985), 117-20.

McGrail, S and Millett, M 1986 'Recovering the Hasholme logboat', *Current Archaeology*, no. 99 (1986), 112-13.

McGrail, S and Switsur, R 1975 'Early British boats and their chronology', *International Journal of Nautical Archaeology and Underwater Exploration*, 4 (1975), 191-200.

McGrail, S and Switsur, R 1979a 'Medieval Logboats of the river Mersey: A Classification Study' in McGrail, S (ed.) *The Archaeology of Medieval Ships and Harbours in Northern Europe: Papers based on those presented to an International Symposium at Bremerhaven in 1979*, British Archaeological Reports international series no. 66 and National Maritime Museum archaeological series no. 5, Oxford, 93-112.

McGrail, S and Switsur, R 1979b 'Medieval Logboats', *Medieval Archaeology*, 23 (1979), 229-31.

Macgregor, M 1976 *Early Celtic Art in North Britain: a study of decorative metalwork from the third century B.C. to the third century A.D.*, Leicester.

McKee, E 1983 *Working Boats of Britain: their shape and purpose*, London.

MacKie, EW 1984 'A late single-piece dug-out canoe from Loch Doon, Ayrshire', *Glasgow Archaeological Journal*, 11 (1984), 132-3.

McInnes, L 1935 *Descriptive catalogue of Kintyre prehistoric antiquities in Campbeltown Museum and of other miscellaneous prehistoric antiquities of the peninsula*, unpublished typescript. Copy held in manuscript collection of NMRS.

M'Intosh, P 1861 *History of Kintyre*, Campbeltown.

MacIvor, I, Thomas, MC and Breeze, DJ 1980 'Excavations on the Antonine Wall fort of Rough Castle, Stirlingshire 1957-61', *PSAS*, 110 (1978-80), 230-85.

Mackintosh, J 1895 *History of the Valley of the Dee*, Aberdeen.

Maclagan, C 1876 'Notes on the Sculptured Caves near Dysart, in Fife, illustrated by Drawings of the Sculptures', *PSAS*, xi (1874-6), 107-20.

MacLean, JP 1923-5 *History of the Island of Mull embracing Description, Climate, Geology, Flora, Fauna, Antiquities, Folk Lore, Superstitions, Traditions with an Account of Its Inhabitants together with A Narrative of Iona The Sacred Isle*, Greenville, Ohio and San Mateo, California, USA.

McNeil, R 1983 'Two 12th-century Wich Houses in Nantwich, Cheshire', *Medieval Archaeology*, xxvii (1983), 40-88.

Macpherson, N 1878 'Notes on Antiquities from the Island of Eigg', *PSAS*, xii (1876-8), 577-97.

Macrae, D 1894 'Notice of a Dish of Bog-butter found at Midton, Inverasdale, Poolewe, Ross-shire, in May 1893', *PSAS*, xxviii (1893-4), 18-19.

Macritchie, D 1890 'Notes on a Finnish Boat Preserved in Edinburgh', *PSAS*, xxiv (1889-90), 353-69.

MacRitchie, D 1896 'Wooden dish lately found in the Hebrides', *The Reliquary and Illustrated Archaeologist*, new series, ii (1896), 241-3.

MacRitchie, D 1912 'The Aberdeen Kayak and its Congeners', *PSAS*, xlvi (1911-12), 213-41.

McVean, DN 1964a 'The forest zone: woodland and scrub' in Burnett, JH (ed.) *The vegetation of Scotland*, Edinburgh, 144-67.

McVean, DN 1964b 'Pre-history and ecological history' in Burnett, JH (ed.) *The vegetation of Scotland*, Edinburgh, 561-7.

McVean, DN 1964c 'Regional patterns of vegetation' in Burnett, JH (ed.) *The vegetation of Scotland*, Edinburgh, 568-78.

Malcolm, J 1910 *The parish of Monifieth in ancient and modern times with a history of the landed estates and lives of eminent men*, Edinburgh.

Mann, LM 1933 'Some Recent Discoveries', *Transactions of the Glasgow Archaeological Society*, new series, viii (1933), 138-51.

Manning, WH 1985 *Catalogue of the Romano-British iron, tools, fittings and weapons in the British Museum*, London.

Mapleton, RJ 1868 'Notice of an Artificial Island in Loch Kielziebar, in a Letter to Mr Stuart, Secretary', *PSAS*, vi (1866-8), 322-4.

Mapleton, RJ 1879 'Notice of the Discovery of an Old Canoe in a Peat-Bog at Oban', *PSAS*, xiii (1878-9), 366-8.

Marsden, PRV 1963 'The Newstead Steering-Oar', *Mariner's Mirror*, 49 (1963), 224.

Marsden, PRV [1967] *A ship of the Roman period, from Blackfriars, in the City of London*, guide publication of the Guildhall Museum, London.

Marsden, P [RV] 1994 *Ships of the port of London: first to eleventh centuries AD*, English Heritage archaeological report no. 3, London.

Martin, C 1992 'Water Transport and the Roman Occupation of North Britain' in Smout, TC (ed.) *Scotland and the sea*, Edinburgh, 1-34.

Marwick, H 1927 'Antiquarian notes on Stronsay', *Proceedings of the Orkney Antiquarian Society*, v (1926-7), 61-83.

Maxwell, HE 1889 'Primitive Implements, Weapons, Ornaments, and Utensils from Wigtownshire', *PSAS*, xxiii (1888-9), 200-32.

Maxwell, S 1951 'Discoveries made in 1934 on King Fergus' Isle and elsewhere in Loch Laggan, Inverness-shire', *PSAS*, lxxxv (1950-1), 160-5.

Maxwell, S 1956 'Paddles from Horizontal Mills', *PSAS*, lxxxviii (1954-6), 231-2.

Megaw, JVS and Simpson, DDA 1979 *Introduction to British Prehistory, from the arrival of Homo Sapiens to the Claudian invasion*, London.

Michie, JG 1910 *Loch Kinnord: its history and antiquities*, revised ed., Aberdeen.

Millett, M and McGrail, S 1987 'The Archaeology of the Hasholme Logboat', *Archaeological Journal*, 144 (1987), 69-155.

Milne, G 1982 'Recording timberwork on the London waterfront' in McGrail, S (ed.) *Woodworking Techniques before A.D. 1500: Papers presented to a Symposium at Greenwich in September, 1980, together with edited discussion*, British Archaeological Reports international series no. 129 and National Maritime Museum archaeological series no. 7, Oxford, 7-23.

Mitchell, A 1864 'On the Vestiges of the Forest of Cree, in Galloway', *PSAS*, v (1862-4), 20-9.

Mitchell, A 1880 *The Past in the Present: What is Civilisation?*, Edinburgh.

Morris, CA 1982 'Aspects of Anglo-Saxon and Anglo-Scandinavian lathe-turning' in McGrail, S (ed.) *Woodworking Techniques before A.D. 1500: Papers presented to a Symposium at Greenwich in September, 1980, together with edited discussion*, British Archaeological Reports international series no. 129 and National Maritime Museum archaeological series no. 7, Oxford, 245-61.

Morris, DB 1892 *The Raised Beaches of the Forth Valley, Stirling*, re-paginated offprint from *Transactions of the Stirling Natural History and Antiquarian Society*, (1892-3), 18-48.

Morrison, A 1980 *Early Man in Britain and Ireland: An Introduction to Palaeolithic and Mesolithic Cultures*, London.

Morrison, I [1975] *Kilmarnock Libraries and Museum, A History, 1797-1975*, printed privately, Kilmarnock.

Morrison, I [A] 1985 *Landscape with lake dwellings: the crannogs of Scotland*, Edinburgh.

Muckelroy, K 1978 *Maritime archaeology*, Cambridge.

Munro, R 1879 'Notice of the Excavation of a Crannog at Lochlee, Tarbolton, Ayrshire', *PSAS*, xiii (1878-9), 175-252.

Munro, R 1880 'Ayrshire Crannogs', *AC*, ii (1880), 17-88.

Munro, R 1882a 'Ayrshire Crannogs (Second Notice)', *AC*, iii (1882), 1-51.

Munro, R 1882b *Ancient Scottish Lake-Dwellings or Crannogs*, Edinburgh.

Munro, R 1882c 'Notes of a Crannog at Friar's Carse, Dumfriesshire', *PSAS*, xvi (1881-2), 73-8.

Munro, R 1885 'The Lake-Dwellings of Wigtonshire', *AC*, v (1885), 73-8.

Munro, R 1890 *The Lake-Dwellings of Europe*, London.

Munro, R 1893 'Notes of Crannogs or Lake dwellings recently discovered in Argyllshire', *PSAS*, xxvii (1892-3), 205-22.

Munro, R 1897 *Prehistoric Problems, being a selection of essays on the evolution of man and other controverted problems in anthropology and archaeology*, Edinburgh.

Munro, R 1898 'The Relation between Archaeology, Chronology and Land Oscillations in post-glacial times; being the opening address to the Antiquarian Section at the Lancaster Meeting', *Archaeological Journal*, lv (1898), 259-85.

Munro, R 1899 *Prehistoric Scotland and its place in European civilisation, being a general introduction to the "County histories of Scotland"*, Edinburgh.

Munro, R 1904 *Archaeology and False Antiquities*, London.

Munro, R 1919 'Notes on Glenfield paddle', *Annals of Kilmarnock Glenfield Ramblers' Society*, 8-9 (1913-19), 62-5.

Munro, R 1921 *Autobiographic sketch of Robert Munro MA MD LLD*, Glasgow.

Murdoch, JB 1882 'Note on a Stone Celt found, in October 1881, on the Estate of Naemoor, the Property of J.J. Moubray, Esq., in the Parish of Muckhart, Kinross-shire', *PSAS*, xvi (1881-2), 430-1.

Murray, D 1898 'A Marine Structure at Dumbuck', *The Scots Pictorial*, iv, no. 83 (29 Oct 1898), 190, 196-7.

Murray, J 1994 'Jade axes from Scotland: a comment on the distribution and supplementary notes', *PPS*, new series, 60 (1994), 97-104.

Murray, P 1902 'Note on a Single-piece Wooden Vessel found in a Peat-moss on Torr Righ Hill, Shiskin, Arran', *PSAS*, xxxvi (1901-2), 582-3.

Murray, SW 1933 *David Murray: a biographical memoir*, Dumbarton.

Name Book (County) *Original Name Books of the Ordnance Survey*. Originals in SRO; microfilm copy in NMRS.

Needham, SP and Longley, D 1981 'Runnymede Bridge' in Milne, G and Hobley, B (eds.) *Waterfront archaeology in Britain and northern Europe*, Council for British Archaeology Research Report series no. 41, London, 48-50.

NMAS 1892 *Catalogue of the National Museum of Antiquities of Scotland*, revised and enlarged ed., Edinburgh.

NMM 1985 *Sea Finland: Finnish seafaring from early history to the future*, guide-catalogue to the exhibition of the same name at the National Maritime Museum, Greenwich, London.

NMRS National Monuments Record of Scotland, The Royal Commission on the Ancient and Historical Monuments of Scotland, John Sinclair House, 16 Bernard Terrace, Edinburgh EH8 9NX.

NSA *The New Statistical Account of Scotland*, Edinburgh, 1845.

O'Drisceoil, DA 1988 'Burnt Mounds; cooking or bathing?', *Antiquity*, 62 (1988), 671-80.

O'Kelly, MJ 1954 'Excavations and experiments in Irish cooking places', *Journal of the Royal Society of Antiquaries of Ireland*, 84 (1954), 105-55.

O'Riordain, SP 1956 *Antiquities of the Irish Countryside*, 3rd ed., London.

O'Sullivan, A 1994 'Harvesting the waters', *Archaeology Ireland*, 8 (1994), 10-12.

O'Sullivan, PE 1974 'Radiocarbon-dating and prehistoric forest clearance on Speyside (East Central Highlands of Scotland)', *PPS*, 40 (1974), 206-8.

Oakley, GE 1973 *Scottish Crannogs*. Unpublished M Phil thesis presented at the University of Newcastle-on-Tyne.

Ogston A 1931 *The prehistoric antiquities of the Howe of Cromar*, Aberdeen.

Orme, BJ 1982 'Prehistoric woodlands and woodworking in the Somerset levels' in McGrail S (ed.) *Woodworking Techniques before A.D. 1500: Papers presented to a Symposium at Greenwich in September, 1980, together with edited discussion*, British Archaeological Reports international series no. 129 and National Maritime Museum archaeological series no. 7, Oxford, 79-94.

OS Ordnance Survey.

Osler, AG 1985 'The North Tyne trows - a logboat link', *Mariner's Mirror*, 71 (1985), 337-8.

Paterson, R 1864 'Note of Human Remains in Wooden Coffins, found in the East Links of Leith', *PSAS*, v (1862-4) 98-100.

Paton, J (ed.) 1890 *Scottish National Memorials*, Glasgow.

Pearson, GW and Stuiver, M 1986 'High-Precision Calibration of the Radiocarbon Time Scale, 500-2500 BC', *Radiocarbon*, 28, no. 2B (1986), 839-62.

Pennant, T 1774-6 *A tour in Scotland and voyage to the Hebrides, MDCCLXXII*, Chester and London.

Peschel, C 1985 'A note on a prehistoric model of a dugout', *International Journal of Nautical Archaeology and Underwater Exploration*, 14 (1985), 265-7.

Petersen, F 1970 'Studies in commemoration of William Greenwell, 1820-1918: Early Bronze Age timber graves and coffin burials on the Yorkshire Wolds', *Yorkshire Archaeological Journal*, 42 (1967-70), 262-70.

Piggott, CM 1953 'Milton Loch Crannog I: A Native House of the 2nd Century AD in Kirkcudbrightshire', *PSAS*, lxxxvii (1952-3) 134-52.

Piggott, S 1938 'The Early Bronze Age in Wessex', *PPS*, new series, iv (1938), 52-106.

Piggott, S 1958 *Scotland before History*, London.

Piggott, S 1983 *The Earliest Wheeled Transport: From the Atlantic Coast to the Caspian Sea*, London.

Plane, AM 1991 'New England's Logboats: four centuries of watercraft', *Bulletin of the Massachusetts Archaeological Society*, 52.1 (spring 1991), 8-17.

Pococke, R 1887 *Tours in Scotland 1747, 1750, 1760*, Kemp, DW (ed.), Edinburgh.

PPS *Proceedings of the Prehistoric Society*.

PSAN *Proceedings of the Society of Antiquaries of Newcastle-upon-Tyne*.

PSAS *Proceedings of the Society of Antiquaries of Scotland*.

Rackham, O 1981 *Trees and Woodland in the British Landscape*, London.

Rackham, O 1982 'The growing and transport of timber and underwood' in McGrail S (ed.) *Woodworking Techniques before A.D. 1500: Papers presented to a Symposium at Greenwich in September, 1980, together with edited discussion*, British Archaeological Reports international series no. 129 and National Maritime Museum archaeological series no. 7, Oxford, 199-218.

Rackham, O 1986 *The History of the Countryside*, London.

Radford, CAR 1959 *The Early Christian and Norse Settlements at Birsay, Orkney*, London.

Ramage, CT 1876 *Drumlanric castle and the Douglasses, with the early history and ancient remains of Durisdeer, Closeburn and Morton*, Dumfries.

Rausing, G 1984 *Prehistoric Boats and Ships of Northwestern Europe: some reflections*, University of Lund, Institute of Archaeology, fran forntid och medeltid no. 8, Lund, Sweden.

RCAHMS The Royal Commission on the Ancient and Historical Monuments of Scotland, John Sinclair House, 16 Bernard Terrace, Edinburgh EH8 9NX.

RCAHMS 1933 *Eleventh report with inventory of monuments and constructions in the counties of Fife, Kinross and*

*Clackmannan*, Edinburgh.
RCAHMS 1956 *An inventory of the ancient and historical monuments of Roxburghshire*, Edinburgh.
RCAHMS 1963 *Peeblesshire: an inventory of the ancient monuments*, Edinburgh.
RCAHMS 1971 *Argyll: an inventory of the ancient monuments, volume i, Kintyre*, Edinburgh.
RCAHMS 1975 *Argyll: an inventory of the ancient monuments, volume ii, Lorn*, Edinburgh.
RCAHMS 1978 *The Archaeological Sites and Monuments of Dumbarton District, Clydebank District, Bearsden and Milngavie District, Strathclyde Region*, Society of Antiquaries of Scotland, Field Survey Project, Archaeological Sites and Monuments Series no. 3, Edinburgh.
RCAHMS 1979 *The Archaeological Sites and Monuments of Stirling District, Central Region*, Society of Antiquaries of Scotland, Field Survey Project, Archaeological Sites and Monuments Series no. 7, Edinburgh.
RCAHMS 1980 *Argyll: an inventory of the ancient monuments, volume iii, Mull, Tiree, Coll and Northern Argyll*, Edinburgh.
RCAHMS 1985a *The Archaeological Sites and Monuments of West Rhins, Wigtown District, Dumfries and Galloway Region*, Archaeological Sites and Monuments Series no. 24, Edinburgh.
RCAHMS 1985b *The Archaeological Sites and Monuments of North Kyle, Kyle and Carrick District, Strathclyde Region*, Archaeological Sites and Monuments Series no. 25, Edinburgh.
RCAHMS 1987 *The Archaeological Sites and Monuments of East Rhins, Wigtown District, Dumfries and Galloway Region*, Archaeological Sites and Monuments Series no. 26, Edinburgh.
RCAHMS 1988 *Argyll: an inventory of the ancient monuments, volume vi, Mid Argyll and Cowal: Prehistoric and Early Historic Monuments*, Edinburgh.
RCAHMS 1992 *Argyll: an inventory of the ancient monuments, volume vii, Mid Argyll and Cowal: Medieval and Later Monuments*, Edinburgh.
Rees, SE 1979 *Agricultural Implements in Prehistoric and Roman Britain*, British Archaeological Reports British series no. 69, Oxford.
Reid, RC 1944 'The Culvennan MSS', *TDGNHAS*, 3rd series, xxiii (1940-4), 41-55.
Renfrew, C (ed.) 1974 *British Prehistory: A New Outline*, London.
Reynolds, DM 1982 'Aspects of later prehistoric timber construction in south-east Scotland' in Harding, DW (ed.) *Later prehistoric settlement in south-east Scotland*, University of Edinburgh, Department of Archaeology, occasional paper no. 8, Edinburgh, 44-56.
Reynolds, PJ 1982 'The Donnerupland Ard' in McGrail S (ed.) *Woodworking Techniques before A.D. 1500: Papers presented to a Symposium at Greenwich in September, 1980, together with edited discussion*, British Archaeological Reports international series no. 129 and National Maritime Museum archaeological series no. 7, Oxford, 129-51.
Riddell, JF 1979 *Clyde Navigation: A History of the Development and Deepening of the River Clyde*, Edinburgh.
Rieck, F and Crumlin-Pedersen, O 1988 *Bade fra Danmarks oldtid*, Roskilde, Denmark.
Ritchie, J 1941 'A Keg of "Bog-Butter" from Skye and its Contents', *PSAS*, lxxv (1940-1), 5-22.
Ritchie, J 1942 'The Lake-Dwelling or Crannog in Eadarloch, Loch Treig: its Traditions and its Construction', *PSAS*, lvvxi (1941-2), 8-78.
RMS Royal Museum of Scotland, 1 Queen Street, Edinburgh EH2 1JD. A constituent body of the National Museums of Scotland: formerly the National Museum of Antiquities of Scotland, and originally the museum of the Society of Antiquaries of Scotland.
Robertson, AS 1954 *The Hunterian Museum: handbook to the cultural collections, including the Roman collections*, guide-publication of the University of Glasgow, Glasgow.
Ross, A 1968 'Shafts, pits, wells - sanctuaries of the Belgic Britons?' in Coles, JM and Simpson, DDA (eds.) *Studies in Ancient Europe: Essays presented to Stuart Piggott*, Leicester, 255-85.
Ross, A and Feachem, RW 1976 'Ritual Rubbish? The Newstead Pits' in Megaw, JVS (ed.) *To illustrate the monuments: Essays on archaeology presented to Stuart Piggott*, London, 229-37.
Russell-White, C and Barber, J 1987 'Burnt Mounds in the East Rhins of Galloway', *Central Excavation Unit and Ancient Monuments Laboratory Annual Report 1987*, Edinburgh, 55-61.
Sale, RM 1981 'Nantwich', *Current Archaeology*, no. 77 (1981), 185-7.
SAS Society of Antiquaries of Scotland. (Manuscript collection housed in the Queen Street, Edinburgh premises of RMS).
Sayce, RU 1945 'Canoes, Coffins and Cooking-troughs', *PSAS*, lxxxix (1944-5), 106-11.
Scott, JG 1976 'The Roman occupation of South-West Scotland from the recall of Agricola to the withdrawal under Trajan', *Glasgow Archaeological Journal*, 4 (1976), 29-44.
Scott, WL 1934 'Excavation of Rudh'an Dunain Cave, Skye', *PSAS*, lxviii (1933-4), 200-23.
Scott, [W] L 1951 'The Colonisation of Scotland in the Second Millennium B.C.', *PPS*, new series, xvii (1951), 16-82.
Scoular, J 1868 'An Investigation of the geological question bearing on the antiquity of the canoes found on the banks of the Clyde', *Transactions of the Glasgow Archaeological Society*, i (1868), 388-90.
Selkirk, R 1983 *The Piercebridge Formula*, Cambridge.
'Senex' 1884 *Glasgow Past and Present*, Glasgow.
Shearer, JE 1907 'Primitive Man and Prehistoric Remains in Britain', *Transactions of the Stirling Natural History and Antiquarian Society*, (1906-7), 93-109.
Sheppard, T 1901 'Notes on the Ancient Model of a Boat and Warrior Crew, found at Roos, in Holderness', *Transactions of the East Riding Antiquarian Society*, ix (1900-1), 62-74.
Sheppard, T 1902 'Additional Note on the Roos Carr Images', *Transactions of the East Riding Antiquarian Society*, x (1901-2), 76-9.
Simpson, DDA 1968 'Food Vessels: associations and chronology' in Coles, JM and Simpson, DDA (eds.) *Studies in Ancient Europe: Essays presented to Stuart Piggott*, Leicester, 197-211.
Simpson, WD 1963 'Sir Daniel Wilson and the Prehistoric Annals of Scotland: a centenary study', *PSAS*, xcvi (1963-4), 1-8.
Simpson, R 1865 *History of Sanquhar*, new ed., Glasgow.
Simpson, R 1895 'Antiquities of Dunscore', *TDGNHAS*, 2nd series, 11 (1894-5), 27-38.
SIMS Scottish Institute of Maritime Studies, Dept. of Scottish History, St Katherine's Lodge, The Scores, St Andrews, Fife KY16 9AJ.
Smith, AG 1981 'The Neolithic' in Simmons, IG and Tooley,

M (eds.) *The Environment in British Prehistory*, London, 125-209.

Smith, J 1895 *Prehistoric Man in Ayrshire*, London.

Smith, WC 1963 'Jade axes from sites in the British Isles', *PPS*, new series, xxix (1963), 133-72.

Smolarek, P 1981 'Ships and ports in Pomorze' in Milne, G and Hobley, B (eds.) *Waterfront archaeology in Britain and northern Europe*, Council for British Archaeology Research Report series no. 41, London, 51-60.

Smout, [T] C 1993 'Woodland history before 1850' in Smout, TC (ed.) *Scotland since Prehistory: Natural Change and Human Impact*, Aberdeen.

*Somerset Levels Papers*. Annual publication of the Somerset Levels Project, Exeter.

SRO Scottish Record Office, HM General Register House, Edinburgh EH1 3YY.

Stevenson, RBK 1981a 'The Museum, its Beginnings and its Development; Part I: to 1858' in Bell, AS (ed.) *The Scottish Antiquarian Tradition: Essays to mark the bicentenary of the Society of Antiquaries of Scotland and its Museum, 1780-1980*, Edinburgh, 31-85.

Stevenson, RBK 1981b 'The Museum, its Beginnings and its Development; Part II: the National Museum to 1954' in Bell, AS (ed.) *The Scottish Antiquarian Tradition: Essays to mark the bicentenary of the Society of Antiquaries of Scotland and its Museum, 1780-1980*, Edinburgh, 142-211.

STUA Scottish Trust for Underwater Archaeology, Department of Archaeology, University of Edinburgh, The Old High School, Infirmary Street, Edinburgh EH1 1LT.

Stuart, J 1854 'On the earlier antiquities of the district of Cromar, in Aberdeenshire', *PSAS*, i (1851-4), 258-63.

Stuart, J 1866a 'Notice of a Group of Artificial Islands in the Loch of Dowalton, Wigtonshire, and of other Artificial Islands or "Crannogs" throughout Scotland', *PSAS*, vi (1864-6), 114-78.

Stuart, J 1866b *Recent Progress of Archaeology: an address given at the opening meeting of the Glasgow Archaeological Society, session 1865-6*, Glasgow.

Stuart, J 1870 'Note on Communication of Lady John Scott, descriptive of Wooden Structures at Spottiswoode, in Berwickshire', *PSAS*, viii (1868-70), 19-20.

Stuart, R 1848 *Views and notices of Glasgow in former times*, Glasgow.

Stuiver, M and Pearson, GW 1986 'High-Precision Calibration of the Radiocarbon Time Scale, AD 1950-500 BC', *Radiocarbon*, 28, no. 2B (1986), 805-38.

Symson, A 1684 'A large description of Galloway the parishes in it' in MacFarlane, W 1906-8 *Geographical Collections Relating to Scotland*, Mitchell, A and Clark, JT (eds.), ii, 51-128.

TDGNHAS *Transactions* (formerly *Transactions and Journal of Proceedings*) *of the Dumfriesshire and Galloway Natural History and Antiquarian Society*.

Timmermann, G 1958 'Zur Typologie der Einbäume', *Offa*, 16 (1957-8), 109-12.

Tipping, R 1994 'The form and fate of Scotland's woodlands', *PSAS*, 124 (1994), 1-54.

Tomalin, DJ 1979 'Barrow Excavation in the Isle of Wight', *Current Archaeology*, no. 68 (1979), 273-6.

Torbrugge, W 1972 'Vor undfruhgeschichtliche Flussfunde. Zur Ordnung und Bestimmung einer Denkmalergruppe', *51.52. Bericht Der Römisch-Germanischen Kommission, 1970-1971*, 1-146.

Twohig, ES 1981 *The Megalithic Art of Western Europe*, Oxford.

Wainwright, FT 1963 *The Souterrains of Southern Pictland*, London.

Walker, NH (ed.) 1980 *A Historical Guide to the County of Kinross: Tours round some places of historical interest in the Shire*, publication of the Kinross Antiquarian Society, Kinross.

Walker, P 1982 'The tools available to the medieval woodworker' in McGrail S (ed.) *Woodworking Techniques before A.D. 1500: Papers presented to a Symposium at Greenwich in September, 1980, together with edited discussion*, British Archaeological Reports international series no. 129 and National Maritime Museum archaeological series no. 7, Oxford, 349-56.

Wallace, PF 1982 'Carpentry in Ireland AD 900-1300 - The Wood Quay evidence' in McGrail S (ed.) *Woodworking Techniques before A.D. 1500: Papers presented to a Symposium at Greenwich in September, 1980, together with edited discussion*, British Archaeological Reports international series no. 129 and National Maritime Museum archaeological series no. 7, Oxford, 263-99.

Wallace, T 1925 'Geology, Archaeology and Early History of Inverness', *Transactions of the Inverness Scientific Society and Field Club*, ix (1918-25), 124-36.

Ward, JE 1974 'Wooden objects uncovered at Branthwaite, Workington, in 1956 and 1971', *Transactions of the Cumberland and Westmorland Antiquarian and Archaeological Society*, new series, lxxiv (1973-4), 18-28.

Warden, AJ 1880-5 *Angus or Forfarshire*, Dundee.

Warner, RB 1986 'The date of the start of Lagore', *Journal of Irish Archaeology*, iii (1985-6), 75-7.

Weeks, J 1982 'Roman carpentry joints: adoption and adaptation' in McGrail, S (ed.) *Woodworking Techniques before A.D. 1500: Papers presented to a Symposium at Greenwich in September, 1980, together with edited discussion*, British Archaeological Reports international series no. 129 and National Maritime Museum archaeological series no. 7, Oxford, 157-68.

Weerd, MJ de 1978 'Ships of the Roman period at Zwammerdam/Nigrum Pullum, Germania Inferior' in Taylor, J du P and Cleere, H (eds.) *Roman shipping and trade: Britain and the Rhine provinces*, Council for British Archaeology research report no. 24, London, 15-21.

Whatley, CA 1987 *The Scottish salt industry, 1570-1850: an economic and social history*, Aberdeen.

Whitaker, I 1954 'The Scottish Kayaks and the "Finn-men" ', *Antiquity*, xxviii (1954), 99-104.

Whitaker, I 1977 'The Scottish Kayaks reconsidered', *Antiquity*, li (1977), 41-5.

Whiting, CE 1937 'Ancient log coffins in Britain', *Transactions of the Architectural and Archaeological Society of Durham and Northumberland*, viii, pt. i (1937), 80-105.

Whittington, G and Edwards, KJ 1994 'Palynology as a predictive tool in archaeology', *PSAS*, 124 (1994), 55-65.

WHM West Highland Museum, Cameron Square, Fort William, Inverness-shire PH33 6AJ.

Wilde, WR 1863 *A descriptive catalogue of the Antiquities in the Museum of the Royal Irish Academy*, vol. i, Dublin.

Wilkinson, TJ and Murphy, PL 1995 *The Archaeology of the Essex Coast, Volume I: The Hullbridge Survey*, (East Anglian Archaeology report no. 71), Chelmsford.

Williams, J 1971 'A Crannog at Loch Arthur, New Abbey', *TDGNHAS*, 3rd series, xlviii (1971), 121-4.

Wilson, D 1851 *The Archaeology and Prehistoric Annals of*

*Scotland*, Edinburgh.

Wilson, D 1862 *Prehistoric Man: Researches into the Origin of Civilisation in the Old and New World*, Cambridge.

Wilson, D 1863 *Prehistoric annals of Scotland*, 2nd ed., Edinburgh.

Wilson, DM 1966 'A medieval boat from Kentmere, Westmorland', *Medieval Archaeology*, 10 (1966), 81-8.

Wilson, G 1862 'Notice of a Crannog at Barhapple Loch, Glenluce, Wigtownshire', *AC*, iii (1882), 52-8.

Wilson, G 1885 'Description of Ancient Forts, etc., in Wigtonshire', *AC*, v (1885), 62-73.

Wood-Martin, WG 1886 *The Lake Dwellings of Ireland: or ancient lacustrine habitations of Erin, commonly called crannogs*, Dublin and London.

Wood-Martin, WG 1895 *Pagan Ireland: an archaeological sketch. A Handbook of Irish Pre-Christian Antiquities*, London.

Wright, EV 1978 'Artefacts from the Boat-site at North Ferriby, Humberside, England', *PPS*, 44 (1978), 187-202.

## Index

(References to gazetteer entries and related illustrations are in bold type)

**Acharacle, logboat (no. 1) 11**, 110, 120, tab. 1
Andersen, S 128
Anderson, J 2
Annandale, ethnographic evidence for logboat use 129
Appleby, Lincs., logboat 134
**Ardgour, bowls (nos. A1–4) 81**, 84, 98, 112, tab. 1
**Arisaig, Loch nan Eala, timbers (no. A5) 81–2**, 112, tab. 1
**Arnmannoch, logboat (no. 2) 11**, 110, 119, tab. 1
**Auchlishie, possible logboat (no. 3) 11**, 110, 121, tab. 1

**Bailemeonach, Mull, trough (no. A6) 82**, 112, 145, tabs. 1 and 11
**Bankhead, mill-paddle (no. A7) 82**, 122, 144, tab. 1
**Barhapple Loch, logboats and paddle (nos. 4–5) 11–12**, 110, 119, tab. 1
**Barnkirk, logboats (no. 6) 12**, 110, 119, 135, tab. 1
Barton, Lancs., logboat 134
**Barry Links, logboat (no. 7) 12**, 110, 121, tab. 1
Beaquoy, Dounby, burnt mound 146
Beverley, Yorks., log-coffin 140
Bishop's Waltham, Hants., log-coffin 140
**Black Loch, logboat (no. 8) 12**, 110, 119, tab. 1
bog butter (and containers) 141, 142, 145, tabs. 10 and 11
**Bowling, logboats and paddle (nos. 9–10) 12–13**, 110, 120, 133, tab. 1
Bowthorpe, Norfolk, log-coffin 140
Branthwaite, Cumbria, logboat 134, 147, 148
Brigg, Lincs., logboat 134
Buchanan, J 117, 126
**Bunloit, Glenurquhart, keg (no. A8) 82**, 122, tab. 1
**Buston, crannog, logboats and oar (nos. 11–13) 13–15**, 110, 119, 123, 125, 126, 133, tab. 1

**Cairngall, log-coffins (no. A9) 83**, 112, tab. 1
**Cairnside, mill-trough (no. A10) 83**, 112, 145, tabs. 1 and 11
**Cambuskenneth, logboat (no. 14)** v, vii, **15–17**, 110, 121, **pl. 1, fig. 2**, tab. 1
Canewdon, Essex, paddle 136

**Carlingwark Loch, logboats (no. 15) 17**, 110, 119, tab. 1
**Carn an Roin, logboat (no. 16) 17–18**, 110, 120, tab. 1
**Carse Loch, logboat (no. 17) 18**, 110, 119, 133, tab. 1
Cartington, Northumberland, log-coffin 140
**Castle Semple Loch, possible logboats (no. 18) 18**, 110, 119, tab. 1
**Castlemilk, possible logboat (no. 19) 18**, 110, 121, 133, tab. 1
**Catherinefield, logboat (no. 20) 18, 20**, 110, 119, 129, tab. 1
Childe, VG 18
**Closeburn, logboat (no. 21)** vii, **19, 20**, 110, 119, 133, 135, **fig. 3**, tab. 1
**Clune Hill, Lochore, logboat (no. 22) 20**, 110, 121, tab. 1
**Cnoc Leathann, Durness, trough (no. A11) 83–4**, 112, 145, tabs. 1 and 11
cooking- and boiling-troughs 145–7
**Craigie Mains, oar (no. A12) 84**, 112, tab. 1
**Craigsglen, logboat (no. 23)** vii, **19, 21**, 110, 121, 133, **fig. 4**, tab. 1
crannogs, chronology and association with logboats 129, 132
**Croft-na-Caber, logboat (no. 24) 21**, 110, 120, tab. 1
**Cunnister, Yell, Shetland, trough (no. A13) 84**, 122, 145, tabs. 1 and 11
Curraghtarsna, Co. Tipperary, burnt mound and possible logboat 147

**Dalmarnock, logboat (no. 25)** vii, **21–3**, 110, 120, 133, 135, **fig. 5**, tab. 1
**Dalmuir, logboat (no. 26) 22**, 110, 120, tab. 1
**Dalvaird Moss, Glenluce, bowl (no. A14)** 81, **84**, 112, tab. 1
**Daviot, timber (no. A15)** viii, 79, **84–5**, 112, **fig. 32**, tab. 1
**Dernaglar Loch, logboat (no. 27) 22**, 110, 119, 133, 135, tab. 1
Dervaird, burnt mound 146
**Dingwall, logboat (no. 28) 22, 24**, 110, 120, 132, tab. 1
Disgwylfa Fawr, Dyfed, log-coffin 140

**Dowalton Loch, crannogs, logboats and bronze vessels (nos. 29–33)** v, **24–5**, 110, 119, 125, 129, 133, 134, **pl. 2**, tab. 1
**Drumcoltran, timber (no. A16)** 85, 112, tab. 1
**Drumduan, logboat (no. 34)** **25–6**, 110, 121, tab. 1
**Dumbuck, logboat (no. 35)** v, vii, **26–7**, 110, 120, 126, 132, **pl. 3, fig. 7**, tab. 1
**Dumglow, log-coffin (no. A17)** 85, 112, tab. 1

**Eadarloch, crannog and possible logboat (no. 36)** vii, **23, 28**, 110, 120, 133, **fig. 6**, tab. 1
**Eadarloch, trough (no. A18)** viii, **85–6**, 97, 112, 145, **fig. 43**, tabs. 1 and 11
Earwood, C 5
**Easter Oakenhead, boat (no. A19)** 86, 112, tab. 1
**Edinburgh, Castlehill, log-coffins (no. A20)** 86, 112, tab. 1
**Edinburgh, Royal Museum of Scotland, timber artifacts (nos. A21–2)** 86, 112, tab. 1
**Errol, logboats (nos. 37–8)** vii, **28–30**, 78, 110, 121, 125, 129, 133, 134, **fig. 8**, tab. 1
**Erskine, logboats (nos. 39–44)** v, vii, **30–2**, 33, 34, 110, 120, 125, 126, 129, 133, **pls. 4–5, fig. 9**, tab. 1

Fahee South, Co. Clare, burnt mound 146
**Falkirk, logboat (no. 45)** **32**, 110, 121, 132, tab. 1
Fenton, A 3, 4
**Finlaystone, logboat (no. 46)** **32**, 110, 120, tab. 1
**Flanders Moss, possible logboat (no. 47)** **32**, 110, 121, tab. 1
**Forfar, logboats (nos. 48–9)** **32, 34**, 110, 121, 129, tab. 1
**Friarton, logboat (no. 50)** **34–5**, 78, 110, 121, 132, tab. 1
*fulachta fiadh* 132, 145

**Garmouth, logboat (no. 51)** vii, **35–6**, 110, 121, 133, **fig. 10**, tab. 1
**Gartcosh House, logboat (no. 52)** **36**, 110, 121, tab. 1
Giggleswick Tarn, Yorks., logboat 128
**Glasgow, Bankton, boat (no. A23)** 86–7, 112, tab. 1
**Glasgow, Clydehaugh, logboats (nos. 53–7)** v, **36–8**, 110, 120, 126, 133, **pl. 6**, tab. 1
**Glasgow, Drygate Street, logboat (no. 58)** **38**, 110, 120, tab. 1
**Glasgow, Hutchesontown Bridge, logboat (no. 59)** v, vii, **38–9**, 110, 120, 133, **pl. 7, fig. 11**, tab. 1
**Glasgow, London Road, logboat (no. 60)** **40**, 110, 120, 132, tab. 1
**Glasgow, Old St Enoch's Church, logboat (no. 61)** **40**, 110, 120, 132, tab. 1
**Glasgow, Point House, logboat (no. 62)** **40**, 110, 120, 133, tab. 1
**Glasgow, Rutherglen Bridge, logboat (no. 63)** vii, **39, 40**, 110, 120, **fig. 12**, tab. 1
**Glasgow, Springfield, logboats (nos. 64–8)** v, vii, **40–4**, 110, 120, 132, 134, 135, **pl. 8, figs. 13–15**, tab. 1

**Glasgow, Stobcross, logboat (no. 69)** **44**, 111, 120, tab. 1
**Glasgow, Stockwell, logboat (no. 70)** **44**, 111, 120, tab. 1
**Glasgow, Tontine, logboat (no. 71)** **44**, 111, 120, 133, tab. 1
**Glasgow, Yoker, logboats (nos. 72–3)** **44–5**, 111, 120, 132, tab. 1
Glastonbury, Som., lake-village 142–4, 147, tab. 12
**Gleann Geal, keg (no. A24)** 87, 112, 145, tabs. 1 and 11
**Glenfield, Kilmarnock, ard (no. A25)** viii, **87–8**, 112, **fig. 33**, tab. 1
**Gordon Castle, logboat (no. 74)** **45**, 111, 121, tab. 1
Gristhorpe, Yorks., log-coffin 140
**Gutcher, Yell, mill-paddle (no. A26)** 89, 112, 144, tab. 1

Hardham, Sussex, logboat 134
Hasholme, Yorks., logboat 122–3, 125, 134, 137
Hove, Sussex, log-coffin 140–1

**Inverlochy, Fort William, timbers (no. A27)** 89, 112, tab. 1
Ireland, ethnographic and historical evidence for logboat use 128–9

Joncheray, D 123–4

Kentmere, Cumbria, logboat 128, 134, 137
**Kilbirnie Loch, logboats (nos. 75–8)** vii, **45–7**, 111, 119, 126, 132, 133, 135, **figs. 16–17**, tab. 1
**Kilblain, logboats and paddle (nos. 79–80)** **47**, 111, 119, tab. 1
**Kilmaluag, Skye, keg (no. A28)** 89–90, 112, 145, tabs. 1 and 11
**Kinross, logboat (no. 81)** **47**, 111, 121, tab. 1
**Kirkbog, paddle (no. A29)** 90, 112, tab. 1
**Kirkmahoe, logboat (no. 82)** **48**, 111, 119, tab. 1
**Knaven, logboat (no. 83)** **48**, 111, 121, tab. 1
**Kyleakin, Skye, keg (no. A30)** 90, 112, 145, tabs. 1 and 11

**Landis, timbers (no. A31)** 90, 112, tab. 1
**Larg, logboat (no. 84)** **48**, 111, 119, tab. 1
**Lea Shun, possible logboat (no. 85)** v, **48–9**, 111, 121, 123, 126, 135, **pl. 9**, tab. 1
**Lendrick Muir, possible logboat (no. 86)** **49**, 111, 121, tab. 1
Liddle, Isbister, burnt mound 146
lime, properties of timber 109
**Lindores, logboats (nos. 87–8)** **49–50**, 78, 111, 121, tab. 1
**Linlithgow, Sheriff Court-house (no. 89)** **50**, 111, 122, tab. 1
**Littlehill, logboat (no. 90)** vii, **50, 56**, 111, 121, 134, **fig. 19**, tab. 1

**Loch a' Ghlinne Bhig, Skye, bowl** (no. A32) **90–1**, 106, 112, tab. 1
**Loch Ard, logboat** (no. 91) **50**, 111, 120, tab. 1
**Loch Arthur, logboats and possible paddle** (nos. 92–3) v–vi, vii, **50–4**, 111, 119, 129, 134, 135, **pls. 10–14, fig. 18**, tab. 1
**Loch Chaluim Chille, logboats** (nos. 94–5) **55**, 111, 120, 135, tab. 1
**Loch Coille-Bharr, paddle** (no. A33) **91**, 112, 138, tabs. 1 and 9
**Loch Doon, boat fragments** (no. A34) **91**, 112, tab. 1
**Loch Doon, logboats and possible paddle** (nos. 96–101) viii, **55–8**, 111, 119, 129, 134, 135, **figs. 20–1**, tab. 1
**Loch Eport, North Uist, trough** (no. A35) **92**, 112, tab. 1
**Loch Glashan, crannog, paddle, troughs, bowls and timber artifacts** (nos. A37–44) vi, viii, **92–8**, 112–13, 138, 145, 147, **pls. 20–2, figs. 38–9 and 44–6**, tabs. 1, 9, 11 and 12
**Loch Glashan, logboats** (nos. 102–3) vi, viii, **58–61**, 111, 120, 134, 135, **pls. 15–17, fig. 22**, tab. 1
**Loch Glashan, possible paddle or oar** (no. A36) viii, 88, **92**, 112, 138, **fig. 34**, tabs. 1 and 9
**Loch Kinellan, crannog and logboat** (no. 104) **60**, 111, 120, 132, tab. 1
**Loch Kinord, boat** (no. A45) **96–7**, 113, tab. 1
**Loch Kinord, logboats** (nos. 105–8) **60, 62**, 111, 120, 126, 134, tab. 1
**Loch Kinord, paddle** (no. A46) viii, 95, **97–8**, 113, 138, **fig. 40**, tabs. 1 and 9
**Loch Laggan, boat** (no. A48) 62, **99**, 113, tab. 1
**Loch Laggan, bowl** (no. A47) 81, **98–9**, 113, tab. 1
**Loch Laggan, logboats** (nos. 109–15) vi, **62–5**, 111, 120, 123, 126, 134, **pl. 18**, tab. 1
**Loch Leven, logboat** (no. 116) viii, 61, **65**, 111, 122, 134, **fig. 23**, tab. 1
**Loch nam Miol, logboats and boats** (no. 117) **65**, 111, 120, tab. 1
**Loch of Kinnordy, logboat** (no. 118) vi, viii, **65–8**, 111, 121, 129, **pl. 19, fig. 24**, tab. 1
**Loch of Leys, logboats** (nos. 119–20) **68**, 111, 121, tab. 1
**Loch of the Clans, logboat** (no. 121) **68**, 111, 121, tab. 1
**Loch Urr, logboats** (no. 122) **68**, 111, 119, tab. 1
**Lochar Moss, logboat** (no. 123) **68**, 111, 119, tab. 1
**Lochar Moss, shoulder-yoke** (no. A49) **99**, 113, tab. 1
**Lochlea (crannog), 'boat'** (no. A50) **99**, 113, tab. 1
**Lochlea, crannog, 'double-paddle'** (no. A52) viii, 88, **99–100**, 113, 138, **fig. 37**, tabs. 1 and 9
**Lochlea, crannog, logboats, oar and paddle** (nos. 124–8) viii, 67, **69**, 111, 119, 132, 134, **fig. 25**, tab. 1
**Lochlea, crannog, oars** (no. A51) viii, 88, **99**, 113, **fig. 35**, tab. 1
**Lochlea, crannog, trough** (no. A53) **100**, 113, 145, tabs. 1 and 11
**Lochlundie Moss, logboat** (no. 129) **69–70**, 111, 121, tab. 1
**Lochmaben, Castle Loch, logboats** (nos. 130–1) viii, **70–1**, 78, 111, 119, 125, **fig. 26**, tab. 1
**Lochmaben, Kirk Loch, logboats** (nos. 132–3) viii, **70–2**, 111, 119, **figs. 27–8**, tab. 1
**Lochspouts, logboat** (no. 134) **72**, 111, 119, tab. 1
logboats: chronology 126–35, tab. 8
logboats: distribution vii, 9, 116–22, fig. 1
logboats: ethnographic and historical evidence 128–9
logboats: forms, size, morphology and indices 5–6, 116, 123–5, 128, tabs. 6 and 7
logboats: identification criteria 1
logboats: preservation and surviving remains 122, 123, 124, 125, tabs. 2, 3 and 4
logboats: propulsion 136–7
logboats: repairs 125–6
logboats: timber types and supply 109, 113–16, 126, tab. 5
log-coffins 137, 139–41
Lucas, AT 128

**Mabie, logboat** (no. 135) **72–3**, 111, 119, tab. 1
**Midtown, trough** (no. A54) **100**, 113, 145, tabs. 1 and 11
mill-paddles and troughs 144–5
**Milton Island, logboat and possible paddle** (no. 136) **72–3**, 111, 120, tab. 1
**Milton Loch, crannog and logboat** (no. 137) **73**, 111, 119, 132, tab. 1
Mitchell, A 2, 3–4
**Monkshill, logboat** (no. 138) **73**, 111, 121, 134, tab. 1
**Morton, logboat** (no. 139) **73–4**, 111, 119, tab. 1
Munro, R 118

Newbarn Down, Isle of Wight, log-coffin 141
**Newstead, steering-oar** (no. A55) **101**, 113, 138, tabs. 1 and 9
North Ferriby, Yorks., paddles 136–7
Northumbria, ethnographic evidence for logboat use 129

oak, properties of timber 109, 113
**Oakbank, crannog, paddle** (no. A56) vi, **101–2**, 113, 138, **pl. 23**, tabs. 1 and 9
oars 136–7
**Oban, log-coffin** (no. A57) **102–3**, 113, tab. 1
**'Orkney', logboat** (no. 140) viii, **74–5**, 112, 122, 134, 135, **fig. 29**, tab. 1

paddles 136–7
**Parkfergus, logboat** (no. 141) **75**, 112, 120, tab. 1
**Perth, Saint John Street, boat** (no. A58) 78, **103**, 113, tab. 1
**Piltanton Burn, timber** (no. A59) **103**, 113, tab. 1
**Plockton, keg** (no. A60) **103**, 113, 145, tabs. 1 and 11
**Polloch River, Loch Shiel, dish** (no. A61) **103–4**, 113, tab. 1
Poole, Dorset, logboat 135

Port Laing, logboats (nos. 142–3) **75**, 112, 121, tab. 1
Portbane, logboat (no. 144) 21, **75**, 112, 120, tab. 1
Portnellan Island, possible logboat (no. 145) **76**, 112, 120, tab. 1

Quernmore, Lancs., log-coffin 141

Ravenstone Moss, paddles (no. A62) viii, 95, **104**, 113, 138, **fig. 41**, tabs. 1 and 9
Redkirk Point, possible logboats (nos. 146–7) **76**, 112, 119, tab. 1
River Arnol, Stornoway, Lewis, dish (no. A63) **104**, 113, tab. 1
River Carron, logboat (no. 148) **76**, 112, 121, 132, 134, tab. 1
'River Clyde', logboat (no. 149) viii, **76–7**, 112, 120, 125, 134, **fig. 30**, tab. 1
River Forth, logboat (no. 150) **77–8**, 112, 121, tab. 1
River Mersey, Warrington, logboats 128, 130, 135, tab. 8
'River Tay', possible logboat (no. 151) **78**, 112, 121, tab. 1
Rough Castle, possible mill-paddle (no. A64) **104–5**, 113, 138, tabs. 1 and 9
Rubh' an Dunain, Skye, possible oar or paddle (no. A65) viii, 95, **105**, 113, 138, **fig. 42**, tabs. 1 and 9

salt-making 147–8

Scott, WL 118
sledges 148
Sleepless Inch, logboat (no. 152) **78**, 112, 121, tab. 1
Staura Cottage, Shetland, paddle (no. A66) **105**, 113, tab. 1
Stirling, King Street, logboat (no. 153) **78**, 112, 121, tab. 1
Stornoway, Lewis, trough (no. A67) **105–6**, 113, tab. 1
Stuart, J 117–18
Stuntney, Cambs., 'tub' 141–2

Talisker Moor, Skye, bowl (no. A68) 81, 91, **106**, 113, tab. 1
Tentsmuir, paddle (no. A69) viii, 88, **106**, 113, 138, **fig. 36**, tabs. 1 and 8
timber vessels, distribution 141–2, tab. 10
Torr Righ Mor, Arran, trough (no. A70) **106–7**, 113, tab. 1
'trow' 129
Tybrind Vig, Denmark, logboats 128, 136

Walthamstow, Essex, logboat 135
Wester Ross, ethnographic evidence for logboat use 129
White Loch, logboat (no. 154) viii, **78–9**, 112, 119, **fig. 31**, tab. 1
Wiesbaden-Erbenheim, Germany, model logboat 129
Williamston, log-coffin (no. A71) **107**, 113, tab. 1
Wilson, D 126